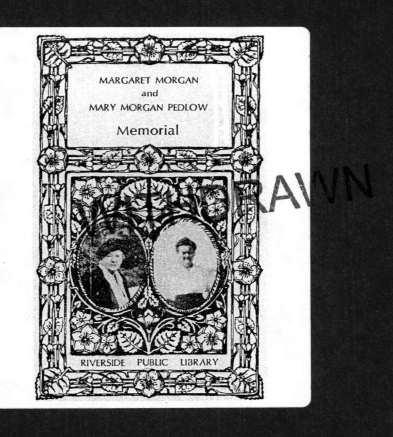

EXPLAINING
the
Universe

★ ★ ★

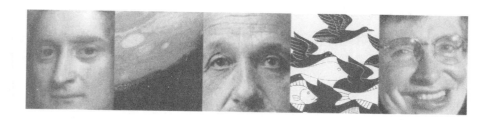

EXPLAINING
the
Universe

The New Age of Physics

★ ★ ★

JOHN M. CHARAP

PRINCETON UNIVERSITY PRESS

PRINCETON AND OXFORD

LIBRARY OF CONGRESS CATALOGING-IN-PUBLICATION DATA
Charap, John M.
Explaining the universe : the new age of physics / John M. Charap.
p. cm.
Includes bibliographical references and indexes.
ISBN 0-691-00663-6
1. Physics. I. Title.
QC21.3 .C48 2002
530—dc21   2001058840

This book has been composed in Goudy
Printed on acid-free paper. ∞

www.pup.princeton.edu

Printed in the United States of America

1 3 5 7 9 10 8 6 4 2

Contents

★ ★ ★

Preface

★ ★ ★

AS THE TWENTY-FIRST CENTURY DAWNED, IT SEEMED TO ME TO BE
a good time to reflect on the state of physics. What are the open questions,
the unresolved puzzles that still command attention? They range from pro-
fundities, such as the structure of space and time, to particular problems—
the safe disposal of nuclear waste, for example. I could not hope to give even
a survey account of all of them. But I do want to convey some indication of
the enormous strides we made in the twentieth century in our understanding
of the physical universe, and of the challenges those advances have themselves opened up. All the claims of science are tentative; therein lies their
strength. For they are not based on unquestioned dogma, which meets the
challenge of opposing argument by increased stridency or deliberate obfusca-
tion. It is rather the ever widening network of support from repeated con-
frontation with experiment, and the internal coherence and consistency of
its laws and their consequences, that give science its resilience and authority.
Of course, it is always possible that we are mistaken, that some fundamental
changes will topple our best-held ideas tomorrow. But even if changes as
profound as those that marked the early years of the twentieth century—

quantum mechanics and relativity—were to make us revise what we presently accept as the foundations of physics, it would not mean that all of our present understanding would have to be abandoned and negated. Newtonian physics has been superseded, but one can still recover it from relativistic quantum physics as an excellent approximation of how the world works, and it still provides a language and a framework for physics at the dawn of the new millennium.

A hundred years ago some scientists thought that there was little left to discover; even now there are some who say something similar. But I believe that the twenty-first century will be no less rich in wonder and discovery than those preceding it. A century ago it was estimated that the universe was around 100 million years old. We now believe it to be more like 13 *billion* years old. In 1900 the smallest entities discussed by physicists were atoms (the existence of which was still disputed by some) and the recently discovered electrons. Electrons were estimated to have dimensions on the order of 10^{-15} meters. Experiments now probe details as small as 10^{-19} meters, and theorists concern themselves with structures as small as 10^{-35} meters. We have now "seen" atoms, broken apart their nuclei; and much of our industry and prosperity is based on the physics of the electron. Techniques developed in research laboratories in the pursuit of "pure" science have a way, sooner or later, of finding applications in new industries and technologies. In the words of the physicist John A. Wheeler: "We live on an island surrounded by a sea of ignorance. As our island of knowledge grows, so does the shore of our ignorance." (Quoted by John Horgan in *The End of Science.* Reading, Mass.: Little, Brown and Company, 1997.) So there are always questions to be asked, problems to be solved. And when we question the universe, we may expect some surprising answers. As J.B.S. Haldane wrote in *Possible Worlds and Other Papers* (London: Chatto & Windus, 1927): "I have no doubt that in reality the future will be vastly more surprising than anything I can imagine. Now my own suspicion is that the universe is not only queerer than we suppose, but queerer than we can suppose."

I ALSO WANT TO MAKE A CASE AGAINST PSEUDOSCIENTIFIC PRATTLE, New Age nonsense, and millennial madness. To read the astrology page in the newspaper may be harmless fun. But to take it seriously is to retreat from light into darkness. Galileo Galilei wrote: "Philosophy is written in that great book which ever lies before our gaze—I mean the universe—but we cannot understand it if we do not first learn the language and grasp the symbols in which it is written." It is not necessary to be a poet to appreciate poetry; neither is it necessary to be a physicist to find wondrous beauty in the discoveries of the last century about the physical universe in which we live. But just as the appreciation of poetry can be sharpened by thoughtful reading, so also with appreciation of physics, there are large rewards for a modest effort of attention. Science is central to our culture; so also is the fictive world of fancy and fantasy. Imagination drives both, but let us not confuse them.

This is not a scholarly work. I have tried to be accurate, but I have not attempted to deal evenly with all the topics that occupy space in the current research literature. Instead, I have selected those which I find most exciting. To set the scene for physics at the end of the twentieth century, I have included a chapter on physics in 1900. In that year, the profound discoveries that led to quantum mechanics and to the theory of relativity lay just below the horizon. What was surely unimagined was the enormous divergence of scale between the very small (now studied as high-energy particle physics) and the very large (astrophysics and cosmology) and the remarkable way that physics today shows them to be intertwined. So I have devoted a considerable portion of the book to these realms. But I don't wish to suggest that all of physics is devoted to these extremes; most research is concerned with phenomena in between, some of which I will explore in chapter 12.

I have benefited from numerous discussions with colleagues here at Queen Mary, and I am grateful for all of them. I would particularly like to record my thanks to Bernard Carr, Jim Emerson, Jerome Gauntlett, Peter Kalmus, and Ian Percival for their helpful comments on some of the chapters. My thanks also to Steve Adams for his help with the illustrations. I have made frequent use of those excellent journals *Physics Today* and *Physics World*, published respectively by the American and the U.K. Institutes of Physics, and have also drawn extensively on the huge diversity of information so freely accessible on the Internet. I have included an index of names, with birth and death dates where available. Trevor Lipscombe and Joe Wisnovsky, my editors at Princeton University Press, have been helpful with criticism and supportive with encouragement, both invaluable for any author.

For each topic I have suggested a few books for further reading, but these lists represent only a small fraction of the many attractive titles intended for the nonspecialist. My hope has been to make some of the physics I find exciting accessible to the general reader. I believe that physicists have a duty to explain to nonscientists what it is that they actually do. In pursuit of this responsibility, I have tried to keep in mind the dictum of a great Russian physicist Yakov Zeldovich: unless you can explain it to a high school student, you don't understand it yourself.

John M. Charap
London, August 2001

A Note on Numbers

★ ★ ★

ALTHOUGH I HAVE SCRUPULOUSLY TRIED TO AVOID MATHEMATICS in this book, I would have found it hard to convey without numbers the scale of phenomena encompassed by physics. Of all the sciences, physics is par excellence concerned with quantitative and precise measurement. Some of the basic physical quantities have been measured to better than one part in a million million, and some measurements agree with what theory predicts to that accuracy. It is hard to comprehend what this precision means; it corresponds to the thickness of a playing card compared with the distance to the moon. When writing either very large or very small numbers, it is far more convenient to use the mathematical notation indicating powers of ten. According to this system of notation, $1,000,000 = 10^6$, and a million million, or $1,000,000,000,000 = 10^{12}$. In other words, a million is $10 \times 10 \times 10 \times 10 \times 10 \times 10$, or the product of *six* factors of ten, and likewise a million million is the product of *twelve* factors of ten. For very small numbers there is a similar notation: one millionth is written as 10^{-6}, which is the result of dividing one by ten six times over. And so on. As well as providing a concise notation, this also facilitates comparison between numbers of very different magnitude. Besides precise determination of relevant quantities, quite a lot of physics is concerned with making order-of-magnitude estimates. As an example, the world's population is now about 6 billion, or 6×10^9; and this is increasing by a staggering quarter of a million, or 2.5×10^5, every day. This increase is about 90 million (9×10^7) a year, which means a proportionate increase by about $(9 \times 10^7) \div (6 \times 10^9) = 1.5 \times 10^{-2}$, or 1.5 percent, every year. Now that's a number I find frightening!

EXPLAINING
the
Universe

★ ★ ★

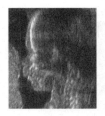

1

INTRODUCTION

★ ★ ★

The Shores of Our Knowledge

THE DISCOVERIES MADE BY PHYSICISTS DURING THE LAST HUNDRED years, and their applications in medicine, industry, and the home, have transformed our lives. We take for granted as everyday necessities what were yesterday only fanciful dreams. Fundamental research has spawned new products and new industries, as well as deeper understanding of the natural world. The pace of discovery and its exploitation is exhilarating and relentless. For all the dangers which sometimes cloud these advances, I find them overwhelmingly positive and enriching.

Think of an expectant mother, attending a hospital clinic today. The position of the unborn child in the uterus is displayed on a screen—ultrasound scanning and computer display instantly combining to give information to the obstetrician and comfort to the mother-to-be (figure 1.1). Her phone call home is transmitted, along with hundreds of others, as laser pulses down an optical fiber thinner than a hair. The child who answers the phone turns off the television—a satellite broadcast—and calls her father. He comes to the

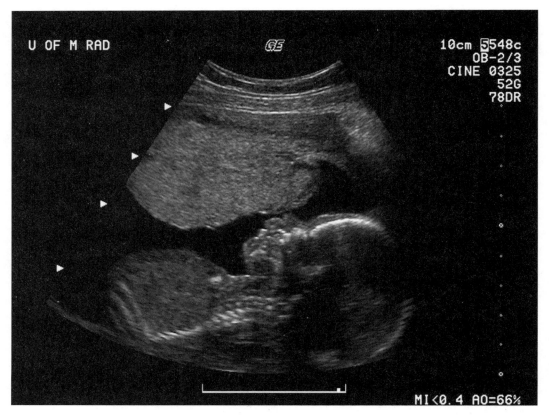

Fig. 1.1. Ultrasound image of an unborn child. Ultrasound imaging is one contribution of twentieth-century physics to medical practice. What future technologies will this child live to see? (Courtesy of GE Medical Systems)

phone, pleased with himself that the evening meal he has prepared can be ready to serve after only a few minutes in the microwave. In the meantime he will listen to a CD, a concert-hall performance recreated by a laser that scans microscopic pits etched in a cheap disc of metal. Nothing very special, yet this family has at its service, and uses with scarcely a thought, wonders that only a lifetime ago would have seemed miraculous.

In this book I have set out to describe the physicist's view of the world at the dawn of the twenty-first century. Most of us enjoy celebrating anniversaries, and we make special note of centenaries. They provide occasions for looking back over the past and also forward to the future. We often find it convenient to package events by centuries, fixing markers in the turbulent flow of events at hundred-year intervals. Fortunately, for this purpose, the dawn of modern physics can be justifiably pinpointed to the year 1900, when Max Planck introduced the law that was to lead to quantum mechanics. A conservative rather than a revolutionary, both by temperament and in his research, he was nevertheless aware that his discovery was of profound importance.[1] As the nineteenth century ended, other seminal discoveries—of the electron and of radioactivity, for example—were made that were also to mark the end of what we now call classical physics.

Before embarking on an account of some of the amazing advances in the twentieth century and looking into the future, it seemed appropriate to survey the prospect from the year 1900. There are people alive today who can recall their world a hundred years ago; and for most of us enchanting mementoes of times gone by, sepia photographs, historic movies, gramophone records, books and journals, can bring that past vividly to our present attention. Some things seem timeless and unchanging; some have vanished forever. But more than the immediately evident change in fashions and style, more even than the changes wrought by our enormously enhanced technological capacity, the last century has seen changes that truly justify the use of the description "revolutionary." These have been most profound in the way that we perceive the world, not least in science, not least in physics.

It is not only through their technological applications that the advances in physics have had such far-reaching consequences. The twentieth century saw profound changes in the way we understand the universe and the laws that codify its structure and content. We learned that the Milky Way, the galaxy of stars of which our sun is just one of a 100 billion others, is itself just one among at least a 100 billion other galaxies. The universe is now known to be not only unimaginably more immense than was conceived just a lifetime ago but also unimaginably older. Its birth in the "big bang" and the fascinating story of its transformation from a terrifying furnace of compressed energy through the successive condensation of matter into stars, planets, and the elements from which we ourselves are made is one of the triumphs of human intellectual elucidation, with mythic resonance the more potent because it is based on fact. In chapter 3 I look back on a century of discoveries in astronomy which have in this way so utterly changed our picture of the universe, and also look to the exciting prospects as new kinds of telescopes open new windows on the cosmos.

It is wonderful enough that matter, in all its variety of forms—from hydrogen, the simplest and most common element, to the most complex molecules, like the DNA that encodes our genetic inheritance—is made from atoms. But those atoms themselves were discovered to have a structure that cannot be properly described or understood without transforming the fundamentals of mechanics as passed on from Galileo and Newton. The invention of this new mechanics—quantum mechanics, the subject of chapter 4—was just one of the revolutionary "paradigm shifts" to punctuate the advance of physics in the twentieth century. The quantum mechanics that is needed to make sense of the structure of atoms has provided a robust underpinning not only for chemistry but also for nuclear physics and the physics of the elementary particles from which all matter is constituted. Today the implications of quantum mechanics for understanding the origin and overall structure of the universe are still being explored. And there are still mysteries to be unraveled in the very foundations of quantum mechanics itself. We know well how to *use* quantum mechanics, but there are still open questions on what quantum mechanics *means*. This is not just a topic for armchair philosophers! For there is a fascinating interplay between information theory and quantum mechanics which gives reason to suppose that we will in the near future have quantum computers far exceeding today's in speed of operation.

But it is not only through the inevitable uncertainties introduced by quantum mechanics that unpredictability enters into physics. Even in the deterministic world of Newtonian mechanics complex systems exhibit chaotic behavior. Chaos is not the same as disorder. There is a strange kind of order in chaos, and how this can emerge is described in chapter 5. The irregularities and discontinuities of chaotic systems need a fresh approach, more holistic than that of classical mechanics. The science of complexity, of the spontaneous generation of order by self-organizing systems, has profound implications for biology, economics, and sociology. It seems that, quite literally, life emerges on the edge of chaos. For it is there that complex adaptive systems flourish—and every living creature, and every ecology, is a complex adaptive system.

Another transformation from the classical worldview inherited from Newton and his rich and fruitful legacy was that wrought by Einstein in his theories of relativity, both special and general. Space and time, respectively the passive arena and ordering principle of classical physics, were unified by the special theory into a spacetime stripped of any absolute landmarks or preferred frame of reference. The general theory then endowed spacetime with a dynamic role as active partner in the dance of matter, and at the same time brought a new understanding of that most familiar of the forces of nature, the force of gravity. The marriage of quantum mechanics with relativity theory required heroic efforts, and even now the full integration of quantum mechanics with the general theory, and not just the special theory, is somewhat speculative and contentious. But undeniably the union of the special theory of relativity with quantum mechanics gave birth to a wonderfully rich and detailed explanation of the physics of elementary particles and the forces between them, an explanation powerful enough to embrace their behavior in high-energy physics laboratories as well as in the high-energy world of the birth and death of stars. For the moment we can only speculate on the fuller implications of quantum theory wedded to general relativity theory.

Those speculations are rich and wonderful. They have convinced most theorists who work at this frontier of physics that the basic entities of the cosmos resemble strings more than pointlike particles, that spacetime has more dimensions than those of which we are aware from our everyday experience, that there are profound interconnections and symmetries that constrain the possible structure of spacetime and matter. To some these ideas may seem to have no more foundation than the imaginings of ancient philosophy or New Age mysticism. But that is not the way I see it! I hope at least to be able to persuade you that they meet the rigorous demands of mathematical self-consistency, and the yet more rigorous demands of conformity with experimental observation. These probing speculations give us not only the dazzle and wonder of an imagined world but the added amazement of knowing that this imagined world may well be the world we actually live in!

The chapter "Your Place or Mine" is about the special theory of relativity and the change in our understanding of elementary particles brought about by bringing it together with quantum mechanics. Relativistic quantum mechanics led to the prediction of antimatter and to the recognition that particles were not permanent, immutable units, like the atoms imagined by the ancient Greeks, but rather could be created or destroyed. From this insight there emerged the idea that particles were best described as packages of energy associated with fields. This is relativistic quantum field theory, perhaps to date the most successful approach to understanding the basic structure of matter and the forces which act on it. Chapter 7 describes how this theory is used to yield amazingly precise agreement with the results of experiments in high-energy particle physics. Quantum field theory is also the setting for what has become known as the standard model of subatomic particle physics, which is the subject of chapter 8.

I have left until chapter 9 some of the developments following from Einstein's general theory of relativity. And in chapter 10, "Strings," I have outlined the fast-developing theory which for the first time has allowed us to bring together quantum mechanics and general relativity, and into the bargain gives a prospect of what has been called a "Theory of Everything." Quantum mechanics is needed to explain what happens on the tiny scale of the atom and its constituents. General relativity is needed to extend Newton's theory of gravity to the extreme conditions of black holes and the grand scale of the universe as a whole. What we now need is a quantum theory of gravity in order to describe the earliest moments after the big bang with which the universe began and so to trace the imprint left from that time on the heavens today. The search for such a theory is what today's cosmology and chapter 11, are about.

Of course there is much more to the physics of today than the exploration by theory and experiment and patient observation of those phenomena at the extremes of size, the microworld of the particle physicist and the macroworld of cosmology. So very much has been learned also on the scales between, the more comfortable scales of our own lives and experience. Many technological advances that have made the new physics possible have also enriched our lives, thus linking the extremes of nature to the everyday. At the dawn of the twentieth century, the then-recent discovery of x rays was already making possible a whole new approach to medicine, an approach that has been extended by subsequent advances in physics to give CT scanners, MRI scanners, ultrasound, and other noninvasive diagnostic tools. Therapeutic advances include laser surgery, radiotherapy and cyclotron beams for cancer treatment, and the whole electronic monitoring technology that has transformed hospital care. But in many ways it has been the advances engendered by the electronic revolution that have had the most pervasive impact on our lives. Transistors, lasers, microwaves, and optical fibers: these are some of the products of twentieth-century physics that have revolutionized our communications, entertainment, and industry. In ancient times, some technologies were developed without any scientific underpinning, based simply on the experience of those who used them. But most of the new modern technologies have been initiated by scientific discovery and driven forward by the ingenuity of the researchers in science laboratories. Whole industries have been spawned by the application of techniques and processes learned from research that was itself motivated by the pursuit of knowledge and understanding for its own sake.

Physics advances on many fronts. High-energy particle physics and astrophysics, glamorous and headline-catching as they are, are not the only fields of excitement and discovery. Even if we may fairly claim to know the fundamental laws governing the behavior of matter in situations less extreme than those of the high-energy laboratory or the outer limits of space and time, it does not follow that we fully understand the implications of those laws. And sometimes, though rarely, a new discovery requires that we revise them. Much more often, we are challenged to explain it within the framework of those laws as we know them.

Let me give an example. From the time he was appointed as director of the laboratory at Leiden in the Netherlands in 1882, Kamerlingh Onnes had sought to push back one of the frontiers of experimental physics: he tried to get closer to the absolute zero of temperature.[2] He was the first to liquefy helium. And in the course of systematic investigation of the optical, magnetic, and electrical properties of substances at low temperatures, he discovered that the electrical resistance of lead suddenly vanished completely at a temperature just 7.2 degrees above absolute zero. He had discovered superconductivity. It was not until forty-five years later, in 1956, that a satisfactory theory was found to explain the phenomenon.[3] But the theory did not require any revision of established laws; rather, it was an imaginative application of them. Superconductors have widespread application—for example, in the magnets used in MRI scanners in hospitals.

The story does not end here, for it was found that certain ceramic materials also lose their electrical resistance when cooled only to the modest extent attainable by using liquid nitrogen.[4] There is still no agreed-upon theory that explains this behavior (the theory that can explain superconductivity in metals such as lead cooled to the much lower temperature of liquid helium doesn't work for these "high-temperature" superconductors). But no one supposes that it will require a revision of the laws governing atomic and molecular structure, still less of the quantum mechanics and electromagnetic theory that underpin them. And though we don't have a theory to explain them, ceramic superconductors are already finding technological applications.

There is a stratified structure in physics which extends to the other sciences. We may believe that at the most fundamental level there are some deep principles and laws that govern all the myriad phenomena of the physical world. But it is unreasonable and unrealistic to start from these fundamentals in order to give a useful explanation for every phenomenon encountered in the laboratory—or in our everyday lives. There can be many steps between the succinct, general, fundamental laws and their complex, specific, and practical applications. What *is* necessary, however, as we move upward through the levels of explanations, of theory and experiment, is that at no step do we find a *contradiction* with what can be deduced from levels deeper down. At any level of explanation it is often useful to introduce what might be called secondary or subsidiary laws, appropriate at that level, encapsulating in a more readily applicable way the consequences of the laws derived from the deeper levels. In this way the pharmacologist, for example, may design a drug using empirical rules about the structure of molecules derived from more basic chemistry; these in turn have a theoretical framework built upon the behavior of electrons and atomic nuclei as governed by quantum mechanics and electromagnetism. But even the theory of quantum electrodynamics is an approximate, albeit effective, theory based on and derivable from a more fundamental level of understanding (figure 1.2).

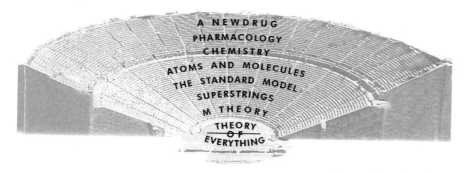

Fig. 1.2. The hierarchy linking basic laws of nature to practical application. There are many steps from a Theory of Everything to the design of a new drug.

The ability to trace back from one level of understanding to a deeper one is related to the *reductionist* approach to science, which is a source of some dismay and conflict, even within science itself. Steven Weinberg, in *Dreams of a Final Theory* (New York, N.Y.: Vintage Books, 1994), writes of "those opponents of reductionism who are appalled by what they feel to be the bleakness of modern science. To whatever extent they and their world can be reduced to a matter of particles or fields and their interactions, they feel diminished by that knowledge. [. . .] The reductionist worldview *is* chilling and impersonal. It has to be accepted as it is, not because we like it, but because that is the way the world works." And to those "scientists who are infuriated to hear it said that their branches of science rest on the deeper laws of elementary particle physics," he replies, "whether or not the *discoveries* of elementary particle physics are useful to all other scientists, the *principles* of elementary particle physics are fundamental to all nature."

One should neither ignore nor belittle the difficulty in moving upward through the strata of levels of understanding. At any step it may be not only convenient but necessary to introduce new laws, new structures, new modes of description, the better to account for the *emergent* phenomena there encountered. It would be not just arrogant, but also stupid, to try to design a drug by starting from quantum electrodynamics! The problems of the pharmacologist are complex and particular; the principles at the deeper level of quantum electrodynamics are simple and general. What is remarkable is that the consequences of these simple, general principles can be so rich and diverse. Quantum electrodynamics may be said to explain the diverse characteristics of the chemical elements, but the wondrous and unique properties of carbon that make it the basis for the chemistry of life must surely be said to be *emergent*. It is rather like an oak tree, in all its complexity, emerging from something as simple as an acorn—but there is a big difference. What emerges from an acorn is always an oak tree, never a zebra. *All* the rich variety of complex phenomena that delight and perplex us arise from the basic laws that physics seeks to determine.

The reductionist may seek for fundamental laws which account for the properties of the particles and fields which are believed to be more fundamental than oak trees or zebras. But that is not to deny that there are important and wonderful laws and regularities—for example, of genetics—which are basic for the understanding of biology. The point is that these are in some sense *consequences* of the chemistry of DNA, and ultimately of the physics of elementary particles. It would be crazy to seek to understand oak trees and zebras are in some ways so alike and in some ways so utterly different by studying elementary particle physics. The chemistry of DNA and the principles of biology that derive from it can be said to *emerge* from the underlying physics of atoms and molecules. And we encounter emergent phenomena within physics itself. The air in your room is made from molecules that collide and scatter from one another in a way well described by mechanics. As they collide, the energy of their motion is distributed among them, and the average kinetic energy accounts for what we call "temperature." The temperature of the air is thus a consequence of molecular dynamics. But it makes no sense whatever to speak of the temperature of a single molecule, nor even of a half a dozen molecules. Temperature is an *emergent* phenomenon, which only becomes significant when a great many molecules are involved.

Unfortunately, the reductionist approach can lead to a misunderstanding that turns some people away from science in general, and physics in particular, in favor of the more comfortable holistic claims of "New Age" beliefs. As one moves up through the levels of phenomena to ever more complex systems, it is not surprising that interesting and important questions are encountered that cannot be answered simply in terms of what is already known. Mysticism and magic have enriched our culture in the past, and science itself should be unashamed to acknowledge its own roots in that subsoil. But science and its offshoots are more reliable and effective aids to solving the urgent problems that confront us than muddy misunderstanding and superstition. Complex problems will often have complex solutions, and there is certainly more to a forest than a collection of trees. But I believe that one can better appreciate the forest as a whole by looking first at the individual trees within it.

More than 80 percent of the scientists who have ever lived are alive today. The pace of scientific discovery has increased and with it the benefits that science brings—and the challenge to ensure that the benefits outweigh the evils and misfortunes that have also arisen from the application of scientific knowledge. I believe it is timely to look back on the past achievements (and the follies) that we so often take for granted. And it is also a good time to take stock of our current picture of the physical world. With the benefit of hindsight, we can see that as the twentieth century dawned, there were premonitory hints of the revolutionary changes that were to come in our understanding of the physical world, changes that have proved necessary to encompass what experiment and observation have revealed. We have had to recast the elegant simplicities of classical physics, but they are still the enduring framework and foundation for the physics of the twenty-first century, and they still provide the context, and even much of the language, within which the subject is advanced. I do not believe that we have come to the end of the road. I do not believe that the modern relativistic quantum view, even when enriched by something like string theory, is the last word. But neither do I believe that we will ever need to completely jettison the theoretical framework we have constructed with so much labor and thought and within which we can fit so much of the subtle and extensive richness of our experimental knowledge. Science thrives on a continued dialogue between explanation and observation. I am convinced that as science spawns new technologies and new techniques, phenomena will be discovered that do not fit comfortably within our present theories. And we should welcome and relish such phenomena. For science can use them as a stimulus to deeper understanding and further advance.

In the following chapters I will try to present the view of the physical world as it now is revealed. The revolutions wrought by the theory of relativity and of quantum mechanics have had profound implications, and they have opened up problems still to be resolved. But relativity and quantum mechanics are now deeply embedded in the model of reality that I have as a physicist, and that I will here try to describe. Yesterday's revolution has become today's orthodoxy, and my views are mainstream, cautious, and, I believe, widely held. Nevertheless, the prospect as I look to the future is challenging and full of deep mysteries.

2

PHYSICS 1900

★ ★ ★

A View from the Past

WHAT WERE THE HOT TOPICS IN PHYSICS IN 1900? AT THAT YEAR'S
meeting of the British Association for the Advancement of Science, Joseph
Larmor (figure 2.1) opened his presidential address to the mathematics and
physics section by emphasizing the advances of the previous twenty years in
the understanding of electricity and magnetism, adding, "In our time the
relations of civilised life have been already perhaps more profoundly altered
than ever before, owing to the establishment of practically instantaneous
electric communication between all parts of the world." The Scottish physicist James Clerk Maxwell had predicted the existence of electromagnetic
waves and identified light as a manifestation of these waves. But what was
the medium that carried them? Larmor asked if it was "merely an impalpable
material atmosphere for the transference of energy by radiation" or rather
"the very essence of all physical actions" (*Nature* 62 [1900]: 449 et seq.).
This controversy over the nature of the "ether" was one of the central questions confronting physics.

Fig. 2.1. Sir Joseph Larmor held the Lucasian Chair of Mathematics of the University of Cambridge. Among those who held the chair before him were Sir Isaac Newton and Charles Babbage; and of those who came after him, Paul Dirac and Stephen Hawking. (© Emilio Segrè Visual Archives, The American Institute of Physics)

Larmor went on to draw attention to progress in the study of electrical discharges in gases, and especially to research on cathode rays. (At the annual meeting of the American Association for the Advancement of Science that year, Vice-President Ernest George Merritt had also emphasized the importance of the study of cathode rays.) Three years earlier, that research had led J. J. Thomson to his discovery of the electron as an elementary constituent of matter.[1] This in turn stimulated an extensive discussion on the structure of atoms. But in Larmor's view "an exhaustive discovery of the intimate nature of the atom is beyond the scope of physics." Nevertheless, speculation on the physics of atoms took up a large part of his address.

In highlighting the controversies surrounding the nature of the ether and the atom, Larmor could not have been aware that he had identified two of the areas of future research that would lead to a radical revision of the basic laws of physics.

At this distance in time, it is hard to recapture the passion with which physicists—and chemists—debated the existence of atoms and molecules. Was one to accept the particulate nature of matter as a reality, or merely a "conventional artifice," useful in explaining experimental facts? Even as the twentieth century dawned some influential voices were heard in opposition to the atomistic model, notably those of Friedrich Ostwald and Ernst Mach. But their views were to be overwhelmed by the irrefutable evidence beginning to emerge from laboratories in the decades leading up to 1900 and beyond. It is noteworthy that one of the crucial steps in the confirmation of the real existence of atoms derives from work of Albert Einstein in his doctoral thesis of 1905, as elaborated in subsequent publications.[2] Interestingly, Einstein was deeply influenced by Mach's ideas on mechanics and the origin of inertia; and it was Ostwald who was the first to propose Einstein for the Nobel Prize.[3]

The nineteenth century had seen the maturation of the physics of electromagnetism. The invention of the chemical battery by the Italian Count Alessandro Volta at the turn of the century initiated several decades of discovery relating to electrical phenomena, progress made possible by the ready availability of sources of electric current. The connections between electricity and magnetism were revealed in experiments made by, among others, André-Marie Ampère and Georg Ohm, who, like Volta, had electrical units named after them.[4] These experiments laid the groundwork for the landmark investigations of that most remarkable of experimentalists, Michael Faraday. He was the protégé of Sir Humphry Davy at the Royal Institution in London. At first he followed Davy's own interests in chemistry, making important discoveries in electrolysis. But by the mid-1830s he had branched out into research areas of his own which led to his discovery of electromagnetic induction, the basis for generation of electricity by a dynamo. It is said that when Mr. Gladstone, the chancellor of the Exchequer, asked Faraday about the practical worth of electricity, he replied, "One day, Sir, you may tax it." Together with the invention of the electric motor by the American Joseph Henry in 1831, Faraday's dynamo laid the foundation of the electric industries, which were soon to transform communications, transport, and manufacture. And by the end of the century electricity was lighting streets and homes.

Faraday's genius was as an experimenter, perhaps the most gifted of all time. He also had a deep insight into the nature of the electric and magnetic phenomena he explored, informed by an imagination that allowed him to make vivid pictorial images of the abstract laws he uncovered. For him space was an all-pervasive ether threaded with lines of force responsible both for magnetic effects, and that transmitting the electric influence of one charge on another. But it was Maxwell who gave these ideas a full mathematical expression, and by postulating an additional contribution to the electric current associated with changing electric fields, even in empty space, he achieved a superb synthesis of electricity and magnetism. Maxwell's equations (1865) are the enduring fruit of this synthesis; they have sensational consequences. For it was immediately apparent to Maxwell that electric and magnetic disturbances propagate through space with a finite speed, and furthermore that an oscillating electric field produces an oscillating magnetic field, and vice versa, so as to allow a self-sustaining *electromagnetic wave* to propagate through space. And he calculated the speed with which it travels. This turned out to be precisely the speed of light. He immediately recognized what this meant: that light must be regarded as an electromagnetic phenomenon. It was not until 1887 that Heinrich Hertz verified the predicted propagation of electromagnetic waves, but by the end of the century Guglielmo Marconi had turned a theorist's predictions to practical application, and in December 1901 succeeded in receiving at St. John's, Newfoundland, signals transmitted across the Atlantic Ocean from Poldhu in Cornwall, England.

Maxwell's unification of electricity, magnetism, and light was the first stride along a path toward a distant goal which today seems to be within our grasp: the bringing together of all the fundamental forces and all the fundamental kinds of matter into one overarching theory.

That light was essentially a wave phenomenon had been accepted since around 1800 and had been reinforced by numerous experimental developments, notably the emergence of interferometry as a precise tool for investigating optical behavior. Interferometry uses the characteristic property of waves that when two or more wave disturbances come together they exhibit interference effects, as here the waves reinforce one another while there they cancel each other out, like ripples on a pond crossing through one another (figure 2.2). The wave nature of light was therefore not in itself a radical prediction of Maxwell's theory. But a puzzle remained, which became increasingly acute. Maxwell had related the speed of light in empty space to fundamental electric and magnetic constants. But the speed of light relative to what? The ether? Earlier in the century, the French physicist Armand Fizeau showed that the speed of light in water was less than that in air, precisely as required by the wave theory. But he also showed that it was different in

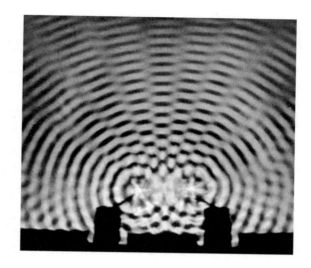

Fig. 2.2. Ripples on the surface of water. They illustrate interference, so characteristic of wave phenomena. (From *Physics*, Physical Sciences Study Committee. D. C. Heath, Boston (1960))

moving water than in water at rest and depended on whether the light was directed along or against the flow of the water. This was not in itself surprising; if you walk along the corridor of a moving train, your speed relative to the track depends on whether you are walking toward the front or the rear of the train. A similar effect would be expected for light passing through moving water. But what about light passing through the ether? The earth moves round the sun at a speed of around 30 kilometers a second, so any terrestrial apparatus would be carried through the ether at sufficient speed to produce a significant difference between the speed of light as measured in the direction of the earth's motion from that at right angles to it. The startling discovery was that no such difference was found, neither in the original experiments conducted by Albert Michelson (the first American physicist to win the Nobel Prize) nor in the later refinements on them that he and Edward Morley conducted in 1887. The resolution of this puzzle would lead in the twentieth century to one of the great revolutions in physics, the theory of relativity, to which we will return in chapters 6 and 9.

Electricity and magnetism and related phenomena were not the only topics mentioned by Larmor. In his address to the British Association, he also made skeptical reference to the methods of statistical mechanics, which make it possible to derive results for the properties of bulk matter from the statistical behavior of its constituent atoms and molecules. In this he was less prescient. Statistical reasoning had already been applied to the kinetic theory of gases, developed by Rudolph Clausius, Maxwell, and Ludwig Boltzmann. This had great success in providing a deeper understanding and extension of the laws of thermodynamics formulated by the French military engineer and mathematical physicist Sadi Carnot, who had been motivated by the very practical concern to improve the efficiency of steam engines. By the end of the nineteenth century, statistical mechanical ideas had also been applied to radiation. In particular, the focus of attention had turned to what is called "black-body" radiation. We are all familiar with the way that as, say, a piece of iron is heated, it begins to glow, at first a dullish red, but as it gets hotter eventually with a bright white light. What determines the intensity (brightness) and color of the light emitted? It could be shown that a perfectly black body, that is to say, one which could absorb any color of light incident on it without scattering or reflecting it back, would also radiate light according to the temperature of the body, but independently of its composition. So the problem could be simplified to the consideration of the intensity and color of the light emitted by an ideal black body. Since this should be given as just some function of the temperature, there was a challenge for thermal physics to arrive at the correct theory. White light is in fact composed from all the colors of the rainbow, so what this really requires is a determination of the intensity of radiation for each color.[5]

It had been shown, using arguments derived from thermodynamics, that there was a single mathematical function, dependent on just one variable, that could express the way that the intensity of radiation at each color depends on temperature. But the derivation of this function had so far evaded the considerable efforts of a multitude of talented theorists. As was said at this same meeting of the British Association, "The investigation, theoretical and experimental, of the form of this function . . . is perhaps the most fundamental and interesting problem now outstanding in the general theory of the relation of radiation to temperature."[6]

The answer to *that* problem, given later on in that very year by Max Planck, was to initiate another of the revolutions that shook the foundations of physics: quantum theory.

And there were other then-recent discoveries that raised puzzling questions. In November 1895, in his laboratory in Wurzburg, Wilhelm Röntgen was studying the electrical conductivity of gas at low pressure. To his surprise he saw a barium platinocyanide-coated screen on the far side of the room glow when he passed a current through his apparatus. By following up this chance observation, he was led to the discovery of x rays, for which in 1901 he was to receive the first-ever Nobel Prize in physics. One of those to whom Röntgen sent copies of the paper announcing his findings was the French physicist Henri Becquerel. The x rays seemed to come from the fluorescent spot where the electron beam struck the wall of the glass tube. Becquerel, who had extensive knowledge of fluorescent crystals, hunted for a crystalline emitter of radiation similar to x rays. Again a chance event, coupled with acute observation and careful investigation, played a role in a major new discovery. For what Becquerel found in March 1896, just five months after Röntgen's discovery, was not a new source of x rays but an entirely new kind of radiation. He had stumbled upon radioactivity, and had thereby opened up a new science, nuclear physics. By 1898 Ernest Rutherford, an 1851 Exhibition scholar from New Zealand working under J. J. Thomson at the Cavendish Laboratory in Cambridge, had discovered that uranium emitted at least two different kinds of radiation, which he called alpha and beta (gamma radiation was reported in 1900 by the Frenchman Paul Villard).[7] The chain of great advances did not stop there, for Marie Sklodowska Curie and her husband, Pierre, were soon to undertake their heroic exploration of the source of Becquerel's radiations and by 1898 had already discovered the new elements polonium and radium. They had founded another new science, nuclear chemistry.

The discoveries of Thomson, Röntgen, Becquerel, and many others had all derived from studies of electrical discharges in gas under reduced pressure. The beautiful, glowing, colorful electrical phenomena in such vacuum tubes were not only on object of study for physicists in their laboratories but a source of parlor entertainment for amateur enthusiasts. The neon advertisements and fluorescent lights we now take for granted had their origin in this technology of the nineteenth century. And it depended on high-vacuum pumps, and the ability to seal metal electrodes in the walls of glass tubes, advances made by Heinrich Geissler around the middle of the century. The

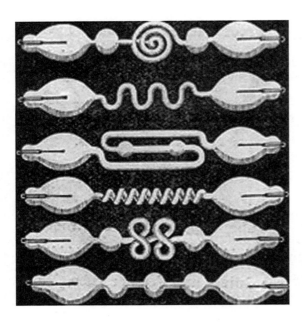

Fig. 2.3. A selection of Geissler's tubes illustrated in a Chicago Laboratory Supply and Scale Company catalog of physical, chemical, and biological apparatus for use in laboratories of colleges, schools, and manufacturers.

Geissler tube did not in itself represent a major scientific advance, but a succession of great scientific discoveries would not have been possible without the inventions of this gifted glass blower (figure 2.3).

Another technological advance that underpinned the great discoveries of the nineteenth century, and which to this day is of crucial importance in laboratory physics, was the use of diffraction gratings in spectroscopy. Newton (figure 2.4) had investigated how a glass prism spreads white light to produce a spectrum with all the colors of the rainbow. The use of diffraction gratings—closely ruled lines on a reflecting surface or on a thin sheet of glass—to analyze spectra had been introduced by Joseph von Fraunhofer, who began his work around 1821. He had investigated and explained the dark lines, now named after him, in the carefully focused spectrum of light from the sun (color plate 1). (They had been observed earlier by William Wollaston.) The dark lines in the spectrum correspond to light *absorbed* from the white light of the sun: they also have their counterpart in the bright lines in the spectrum of light *emitted* by chemicals heated in a flame. The brilliant yellow that common salt gives to the flame of a kitchen gas stove is a familiar example of this phenomenon. Salt is sodium chloride, and a prominent bright yellow line in

Fig. 2.4. Sir Isaac Newton. (Portrait by Murray, reproduced with permission from the Master and Fellows of Trinity College, Cambridge)

the spectrum signals the presence of sodium. The application of these ideas to spectroscopic chemical analysis was largely developed by Gustav Kirchhoff. Something like a chemical fingerprint, the pattern of lines in the spectrum reveals the chemical elements present where the light is emitted or absorbed. And just as forensic experts can use fingerprint analysis to identify those who were present at the scene of a crime, scientists could use the lines in the spectrum to identify the elements responsible for them—and even to discover new ones. In 1868 Sir Joseph Lockyer used a spectroscope adapted to a telescope to obtain a spectrum of light from the sun. He found a yellow line unrelated to any of the elements known at the time and inferred the existence in the sun of a previously unknown element, which he named helium, from the Greek word for the sun, *helios*. It was not until 1895 that helium was isolated by Sir William Ramsay in the "emanations" from uranium (the gas given off in what we now know as radioactive decay).

The precision of grating spectroscopy requires very accurately ruled lines,

Fig. 2.5. Henry Rowland and his ruling engine. (The Ferdinand Hamburger, Jr., Archives of The Johns Hopkins University)

and significant advances followed from Henry A Rowland's development, at the Johns Hopkins University in Baltimore, of a "ruling engine," which by 1882 was able to make gratings with 43,000 lines to the inch (figure 2.5).[8] Gratings have made spectroscopy and related interferometric methods in optics among the most precise techniques in science. And as already illustrated by the discovery of helium, when turned to the heavens, spectroscopic investigation can reveal details of the composition of astronomic objects directly from the light we receive from them. The modern science of astrophysics stems from such observations.

To turn once more to Larmor's address: at its philosophic heart was a discourse on the theoretical developments in mechanics, in particular the reformulation of Newton's laws of motion encapsulated in the analytical mechanics that had been brought to maturity by the Irishman Sir William Hamilton and the German mathematician Carl Jacobi.[9] Newton had formulated his laws in terms of particles acted on by forces; his equations expressed the change in the motion of the particles brought about through the action of these forces. For complicated mechanical systems, composed of many particles, it is difficult to handle these equations directly, and the eighteenth and nineteenth centuries had seen the invention of powerful techniques which permitted a more direct approach.

It turns out that the equations of motion which follow from Newton's laws can also be obtained by taking a single quantity constructed from the energy of the particles and the forces between them, and then insisting that this quantity is at an extremum.[10] This is analogous to the way that one may determine that the configuration of an elastic band stretched around three pegs on a board is a triangle with straight sides, either by working out what the forces are which align each little piece of the elastic, or, far more simply, from the single requirement that the actual configuration is the one with the shortest length. The quantity used in analytical mechanics that is analogous to the length of the elastic is called the *action*; the requirement that it is at an extremum is called the *action principle*.[11] It turns out that the action principle, the theory of "canonical transformations" developed by Jacobi, and a quantity related to the energy introduced by Hamilton (now called the Hamiltonian function) all play an important part in the formulation of what was to replace classical Newtonian mechanics. This was quantum mechanics.

With hindsight, one can say that as the twentieth century dawned, it was already possible to discern precursors to the upheavals that would reshape the foundations of physics in the years to follow. But the prevailing view in the community of physics was an optimistic acceptance that the progress of what we now call classical physics would continue, that new discoveries would fit smoothly into the growing, coherent pattern of observation and theory which had been established with so much labor and ingenuity. Little did anyone foresee what was so soon to follow!

3

HEAVENS ABOVE

★ ★ ★

Of Galaxies and Stars

THOSE OF US WHO LIVE IN CITIES ARE DENIED THE AWESOME SIGHT of a sky full of stars. Undimmed by the pollution and the scattered light from our streets, the night sky is one of nature's most majestic spectacles. On a clear, moonless night it is possible to discern a few thousand stars with the unassisted eye. Even a pair of binoculars or a modest telescope will reveal many times that number. The Milky Way, which to the naked eye is a cloud-like belt of light across the sky, can be resolved into a myriad of stars—our galaxy (plate 2). As the earth turns, the northern heavens appear to revolve around the Pole Star (figure 3.1), making the belief of the ancients in a celestial sphere all but irresistible. The ancients likewise saw the stars as delineating their gods and mythic heroes. Their constellations gave names to the stars which we still use today. In the constellation Andromeda, there is a nebula, which the naked eye can just discern as a smudge of light rather than a pointlike star. This had already been noted over a thousand years ago by the Persian astronomer Azophi Al-Sufi; a millennium later it was to play a

Fig. 3.1. A time exposure of the night sky, showing the trails of the stars as they appear to revolve around the Pole Star. (Anglo-Australian Observatory, photograph by David Malin)

dramatic role in changing our perception of the universe. For the Andromeda Nebula is not just a glowing cloud of gas, such as we find elsewhere in the Galaxy, but can be seen through a telescope to contain many individual stars (plate 3). It needed a giant telescope and a superb astronomer to see that it was not a cluster of stars like the many others already studied but, as we shall see below, had a far greater significance.

For all the apparent constancy of the stars, the heavens do change. The planets, the "wanderers" known from times immemorial, change their positions with respect to the "fixed stars" as they orbit the sun. Comets too, in their circuits through the solar system, blaze into prominence as they approach the sun, and then fade as they recede again to the outer limits of the solar system. Even in the remote reaches of space beyond the solar system, one can observe change. Occasionally a new star will appear, a *nova*; in fact, this is the result of an *old* star, previously too faint to have been visible, flaring into visibility as it draws material from a nearby companion. Still more rare is a *supernova*, such as the one visible to the naked eye in the Southern Hemisphere early in 1987 (plate 4), in which a star is destroyed in an explosion which briefly makes it one hundred thousand million (10^{11}) times brighter than before. There are also stars which grow brighter and then

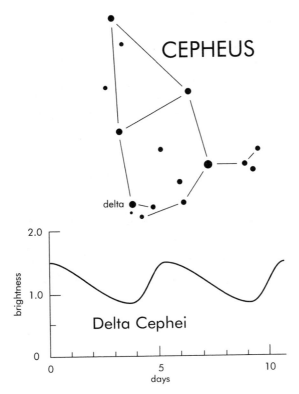

Fig. 3.2. The changing brightness of the prototype Cepheid variable star.

fade in a regularly repeating alternation, like a celestial signal lamp. One class of such stars is called the Cepheids; and the period of the regular pulsations of a Cepheid variable can be as short as a day for some, or as long as about fifty days for others (figure 3.2). What is remarkable is that the period of each star is directly related to its luminosity, its intrinsic brightness.[1] This means that one can deduce the luminosity of a Cepheid, from its periodicity, which is easily measured. The more remote a star, the fainter it will appear, with its apparent brightness reduced by a factor proportional to the square of its distance, the well-known inverse square law. So by observing the apparent brightness of a Cepheid, and having deduced its luminosity from its period, one may determine how far away it is. For at least some of the Cepheids, however, to establish the relation between period and luminosity one needs a more direct method for determining distance. Some of them, including the "original" Cepheid variable, Delta Cephei, are close enough for the surveyor's technique of triangulation to be used, with the diameter of the earth's orbit round the sun as a baseline. Once having established the relation between the period and luminosity from these Cepheids, one is able to be use Cepheids in general as "standard candles" in establishing distance scales in the Cosmos.

The speed of light is 3×10^8 meters per second (or if you prefer, 186,000 miles per second).[2] It takes light about eight minutes to reach us from the sun, which is in fact a very ordinary sort of star. The next nearest star, Proxima Centauri, is about four light-years away, which means it takes four years for its light to reach us. The Milky Way galaxy, in which our sun occupies a modestly suburban position—about two-thirds of the way out from the center—is a disc-shaped aggregate of stars about 10^5 light-years across. It contains some 10^{11} stars: that's about 50 million times more stars than one can see with the naked eye. A hundred years ago it was thought that beyond the Galaxy was nothing but empty space. The Galaxy was the universe.

In 1923 the largest optical telescope in the world was the giant telescope on Mount Wilson in California, with its 100-inch-diameter mirror. Its light-gathering capacity made it a powerful instrument for observing very faint objects in the sky. And in October 1923 Edwin Hubble was using this telescope to search for novae in the Andromeda Nebula (figure 3.3). Nebulae like that in Andromeda were of interest, because it was already known that they contained very many stars; in fact, such galaxies of stars were already being considered as possible "island universes." It was understood that the Milky Way was a stupendous collection of stars—but did the Andromeda Nebula lie within it, or was it perhaps a nearby companion to it?

By making repeated observations of the nebula, Hubble was able to identify some stars that changed in brightness, some novae. But then he recognized that among them were stars with the regular alternations of brightness characteristic of Cepheid variables. Taking together their apparent brightness and the period of their variation, Hubble calculated their distance. They were 3 *million* light-years away. From their enormous distance, it was clear that they were *not* inside our galaxy, already known to be about 10^5 light-years in diameter. He concluded that the Andromeda Nebula is another gal-

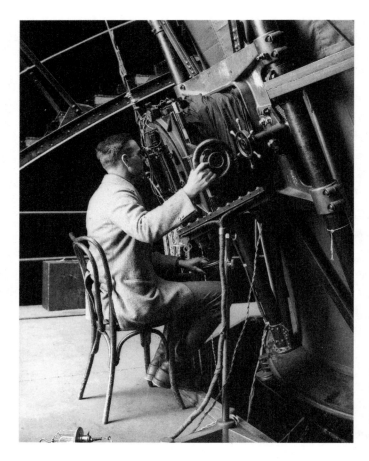

Fig. 3.3. Edwin Hubble at the Hooker 100-inch telescope on Mount Wilson. (The Observatories of the Carnegie Institution of Washington)

axy, very similar to our own, lying well outside the Milky Way. Soon other nebulae were also identified as distant galaxies. But Hubble's greatest contribution to astronomy was yet to come.

The spectra of light from stars show dark lines, the Fraunhofer lines, associated with absorption of light by the hot gases at their surface. As we saw in chapter 2, these line spectra are characteristic of the elements which produce them, and to the trained eye their pattern is a signature of the chemical composition of the gas. The wavelength of light, which is what determines its color, is shifted toward the blue end of the spectrum if the source is moving toward the observer and to the red if it is receding. (A similar shift in wavelength occurs also for sound waves, as had been predicted in 1842 by

Christian Doppler. It is responsible for the familiar change in pitch of the siren of an ambulance as it passes by.) What had already been inferred from this "Doppler shift" of the dark lines was that the Andromeda Nebula was rushing toward us at nearly 300 kilometers per second. Hubble now turned his attention to other galaxies and again determined their distance using the Cepheid-variable method. For more than a decade, Vesto Slipher had been compiling data from the line spectra of galaxies; he found that thirty-six of the forty-one he had studied showed a redshift, which indicated that they were receding. Howard Robertson noted from Hubble's distance measurements and Slipher's redshifts that there appeared to be a relation between them: the speed of recession increased with distance. Hubble refined this analysis to conclude that all of the more distant galaxies were receding from ours. He found that each galaxy's speed of recession was, to good approximation, directly proportional to its distance. This linear relation is *Hubble's Law*, which plays a fundamental role in determining the distance scale in the universe.[3]

Now, you might think that this recession of all the distant galaxies from our own means that we are in a privileged position in the universe. But it is a basic assumption of modern cosmology, dating back to Nicolas Copernicus, that there are no such privileged positions. This premise, called the *cosmological principle*, affirms that on a sufficiently large scale, the universe is at all times homogeneous, every part of it much the same as every other part. Of course, the surface of the earth differs from the center of the sun or from intergalactic space! But on a scale larger than the distance between clusters of galaxies, these differences can be averaged out, and then the universe indeed appears to be homogeneous. If one accepts that there is nothing very special about our own place in the universe, the recession of the galaxies observed by Hubble implies that the galaxies are not just receding from *us* but are all receding from one another. This can best be explained by supposing that the whole universe is expanding, with a corresponding increase in the distances between *all* the galaxies. It is not that the galaxies are rushing away from one another *through* space: space itself is expanding, carrying the galaxies with it. It's rather like watching a muffin bake: the raisins in the muffin get further and further apart as the dough between them expands.

In addition to this grand, overall expansion are the smaller motions related to the gravitational attraction between neighboring galaxies—which is why, for example, the Andromeda galaxy is approaching us rather than receding from us. The expansion of the universe is one of the key facts in cosmology, and the constant that relates the speed of recession to distance in Hubble's observations is one of the most crucial parameters in the quantitative study of cosmology. This constant permits us to calculate the age of the universe and to trace back the expansion to a time when the whole universe was infinitesimally small—to the big bang with which everything began. Despite continuing refinement of Hubble's methods and a vast accumulation of data over many years, the precise value of Hubble's constant is still somewhat uncertain, but it is generally accepted that the speed of recession (in kilometers per second) is close to twenty times the distance (in millions of light-years).[4] On the assumption that the Hubble's law relating redshift to distance is reliable, it is now possible to determine the distance of remote galaxies out to the furthest reaches of observation. Galaxies have been observed with a redshift so large that the wavelength of the light detected has been stretched to as much as six times its original size.

Modern telescopes make visible ever fainter objects in the heavens, probing deeper and deeper into space. But since it takes time for the light to reach us from distant objects, looking out to greater distances also means looking back into the past. The light from a galaxy a billion light-years away has, after all, taken a billion years to reach us! So we see that galaxy not as it is now but as it was a billion years ago. And we do see galaxies at that distance in space and in time—millions of them. Using NASA's appropriately named Hubble Space Telescope, photographs have been taken that look out deeper into space than ever before, capturing images nearly 4 billion times fainter than the limits of unaided human vision (figure 3.4 and plate 5). Although these deep-space surveys only cover small patches of the sky, they are believed to be representative, and from them we may infer that in all directions there are myriads of galaxies as far away as 12 billion light-years. There are *at least* as many galaxies in the universe as there are stars in our own Milky Way.[5]

Almost from the time of Hubble, it was known that our galaxy and the

Fig. 3.4. A view of the Hubble Space Telescope, with the earth below. This photograph was taken during the mission to correct the optics of the telescope. (NASA and NSSDC)

Andromeda galaxy were just two of a group of neighbors, the local cluster, which includes also the Magellanic Clouds, two small nearby galaxies visible to the naked eye in the Southern Hemisphere. Galaxies are often found in clusters, many of them far richer than our own local cluster (plate 6). And surveys conducted over the last two decades have shown that these clusters of galaxies are not the largest structures in the universe; for they are themselves aggregated in superclusters, composed of tens to hundreds of clusters and extending as filaments over hundreds of millions of light-years. Between these strung-out superclusters there are vast voids, enormous spaces containing few, if any, galaxies. We have no evidence of any structure beyond the scale of superclusters. The universe at larger scales appears to be pretty much the same everywhere, uniform and homogeneous. It is only at these stupendous scales that the Copernican cosmological principle comes into play.

Telescopes like the 100-inch and 200-inch reflectors in California's great observatory complex on Mounts Wilson and Palomar or the Hubble Space Telescope, with their superb light-gathering power and precise optics, are not the only tools with which modern astronomers can observe the cosmos.[6] The portion of the spectrum to which our eyes, or indeed our photographic plates, are sensitive is but a small fraction of the range of wavelengths, from long radio waves to extremely short gamma rays, all of which are emitted by astronomical sources of one kind or another, all of which are now studied with the aid of new kinds of telescope.

Inventions first employed in wartime led rapidly in the second half of the twentieth century to the development of radio astronomy. One of the spectral lines associated with electrically neutral hydrogen atoms has a wavelength of 21 centimeters: although this line is very faint, it is important, because hydrogen is the most abundant element in the universe, and astronomers can use it to explore the tenuous gas between the stars. And 21 centimeters fell within the range of wavelengths previously the domain of military radar. So radar-based technology led to devices that could search the heavens for sources of this emission from hydrogen, and over the years these instruments have developed in their capacity to capture radio waves from outer space with wavelengths ranging from meters down to the millimeter waves where radio-based techniques shade off into those adopted by infrared astronomers. From infrared through to the ultraviolet, adaptations of the methods of optical astronomy can be made. A difficulty is that Earth's atmosphere is opaque to some of the wavelengths that astronomers wish to study. The absorption lines of gases in the stratosphere and above, as well as bands of opacity, leave open "windows" on only a portion of what would become observable above the atmosphere (figure 3.5). Telescopes on mountaintops or observations from high-altitude balloons may get one above the clouds, but one is still impeded by water vapor and other bothersome gas in the upper atmosphere. To get above that too required the liberating assistance of artificial satellites, and it is with detectors carried on these that the new branches

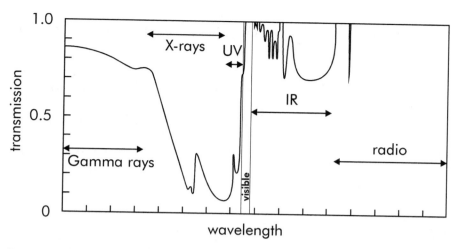

Fig. 3.5. Windows on the cosmos. Water vapor and other gases in the atmosphere absorb radiation from the stars, except for some "windows"—wavelengths in which the absorption is much reduced, so enabling astronomers to look out and explore the cosmos. It is the window extending over the range of wavelengths of visible light that allows us to see the stars.

of astronomy really, and literally, took off. There are now substantial programs for satellite-based astronomy, making observations in the radio, infrared, ultraviolet, x-ray, and gamma-ray portions of the spectrum, as well, of course, as the visible.

In the late 1960s, at the height of the Cold War, the United States placed a series of satellites in orbit to monitor possible violations of the Nuclear Test Ban Treaty. They were equipped with detectors designed to respond to the gamma rays which would be emitted by a nuclear explosion in space. And they *did* detect bursts of gamma rays; but it soon became apparent that they had nothing to do with nuclear weapons but came rather from outer space. For several years the existence of these mysterious gamma-ray bursters, as they have become known, was a military secret, but in 1973 the U.S. government declassified the information on their observation, and with the addition of research satellite data in the 1990s they became one of the hottest topics in astrophysics. These bursts of gamma rays are observed about once a day; they appear throughout the sky and last for anything from less than a second to several minutes. Because they are distributed very uniformly across the sky, it became apparent that their source is not to be found in the Galaxy, although it was considered possible that they might originate in the galactic "halo," the diffuse, spherical cloud of stars which extends about a million light-years beyond the main disclike concentration of stars which make up the Milky Way. They are surprisingly bright, so that if they originate outside the Galaxy, they must be *extremely* powerful sources of radiation. In the first 10 seconds a burster emits more energy than has been radiated by the sun in its 10 billion years of existence! Their nature and location remained controversial until 1998.

A satellite called BeppoSAX, carrying both gamma-ray and x-ray detectors, established that after the gamma-ray flare-up from a burster has died down, there remains a glow of x rays that may continue for some days. Two such observations in 1997 were to lead to spectacular results. On 28 February the satellite detectors recorded a gamma-ray burst; within a few hours the project team had pointed the x-ray telescope in the direction of the burst and detected a rapidly fading x-ray source. This was also seen by the Ulysses satellite, which has as its prime mission the study of the sun. Then astronomers using the British-Dutch telescope at La Palma were able to record an optical image of the location of the burst within less than a day from its first discovery. Another image of the same patch of sky taken eight days later showed that a light source visible in the first optical image had faded by the time the second was made—this fading star was presumably visible light coming from the burster. The Hubble Space Telescope then found a faint galaxy in the same part of the sky (plate 7); perhaps the source of the burster was in that galaxy. Ten weeks later another burster was tracked in a similar way, but this time it was possible for the optical observations to be made with the 10-meter Keck II telescope in Hawaii, and at last an optical spectrum was obtained. The Doppler shifts in the spectral lines could then be used to infer the distance to the source of the burst. It is at least a staggering 7 billion light-years away, one of the most distant objects ever studied. This means that it had to be also an incredibly powerful source of radiated energy for it to appear as bright as it did at such an extravagant distance. It is now generally accepted that gamma-ray bursts come from immensely distant galaxies, and that their prodigious outpouring of energy may originate from collisions between binary neutron stars.

The astronomy of the past fifty years has exposed a host of unexpected wonders. Amongst them are *quasars* (quasi-stellar radio sources), immensely powerful radio sources emitting enormous amounts of energy from a region much smaller than the Galaxy. It is now generally believed that each is associated with a giant black hole in the core of a galaxy; and it is even speculated that *every* galaxy, including our own, has a black hole at its heart, although not all of them are in an "active" phase, when prodigious amounts of energy are released as stars fall into them.[7] Most quasars are billions of light-years away, which means that they are very old. Quasars may not be restricted to the early stages of the universe, and it has been suggested that there is a quasar embedded in the very bright radio source Cygnus A, which is relatively nearby, being "only" 600 million light-years away. Some quasars also emit gamma rays, and one of these flared into brilliance in June 1991, to become one of the brightest high-energy gamma-ray objects in the sky—especially extraordinary considering its tremendous distance of about 4 billion light-years. It is an example of a new class of active galaxies known as gamma-ray blazars. Blazars are a class of quasars which exhibit very rapid variability in intensity, thought to arise from the emission at close to the speed of light of a jet of material pointed toward us, generated by a supermassive black hole. A network of observers working in concert—called the Whole Earth Blazar (WEB) Telescope—has been formed to provide continuous monitoring of the sky for their dramatic but short-lived flare and fade.

Black holes more massive than a million stars may be found at the heart of most galaxies. But it is not only such supermassive black holes which are found in the heavens. To become a black hole may be the ultimate fate of quite ordinary stars. We know a great deal about how stars are born and how they die; understanding the life cycle of stars is one of the major achievements of twentieth-century astrophysics. Stars are born when diffuse clouds of hydrogen and traces of other material condense through their own gravitational attraction, becoming hotter as they do so (like the air compressed in a bicycle pump). Eventually, if the collapsing cloud becomes hot and dense enough, thermonuclear reactions are ignited which, through a complicated cycle of changes, turn the hydrogen into helium and eventually into more massive elements (plate 8). The process by which such concentrations of gas collapse is well enough understood so as to explain how stars can be formed with masses upward from around 8 percent of the mass of the sun (below that mass, the gas never gets hot enough for thermonuclear reactions to take place). What happens then is again complicated but by now well understood. How the star's life ends depends on its mass.

Although Sir Arthur Eddington had considered the possibility that a star might collapse to become a black hole, he had rejected the idea as too ugly and implausible. It was J. Robert Oppenheimer and Hartland Snyder who, in 1939, first took the idea seriously and deduced that a sufficiently massive star was likely to evolve into a black hole. And there is by now ample evidence that some of the objects discovered, especially some of the x-ray sources, are associated with matter falling into such stellar-remnant black holes.

A more common outcome of stellar evolution is for the star to become a *white dwarf*, and there are many of these known to astronomy. These are stars that have collapsed under their own weight to become the size of Earth, all their fuel spent, but still glowing white-hot—a cooling ember of the thermonuclear fire that made them shine in an earlier phase of their evolution.

Yet another form of collapsed star is only a few kilometers across, with matter compressed to the density of an atomic nucleus. These are the neutron stars, some of which are manifest as *pulsars*, flashing like celestial lighthouses as they spin round, regularly beaming a shaft of radiation toward us. The first pulsar to be discovered was found by Jocelyn Bell, who was then a research student at the University of Cambridge, working with the radio astronomer Antony Hewish. She was looking for scintillations in the radio signals from stars, analogous to the familiar "twinkling optical." Using the high-resolution radio telescope Hewish had built, an array of 2048 antennae covering an area of 18,000 square meters, she noted that one of the sources she was observing seemed to be not just scintillating (twinkling, like the stars we see with our own eyes) but emitting extraordinarily regular pulses. At first it seemed incredible that such regularity could originate without intelligent intervention;[8] but there was a theoretical prediction that could explain just such a phenomenon, and after eliminating alterative explanations, astronomers now agree that Bell's pulsar is indeed a rapidly rotating neutron star. In 1974 Hewish shared the Nobel Prize for its discovery, the first ever awarded for a branch of astronomy. Jocelyn Bell did not share that prize; but many physicists agree that she should have.[9]

In the constellation Taurus, Chinese astronomers in 1054 observed a "guest star," which we now know was a supernova; it blasted out the cloud of gas still visible as its glowing shroud. This is the Crab Nebula; at its heart there remains a faint star, which is also a pulsar (plate 9). There are many other expanding clouds of gas surrounding remnants of supernova explosions. The supernova explosion observed in February 1987 in our companion galaxy, the Large Magellanic Cloud, is still the subject of intensive observation. It is expected that its remnant will become a pulsar, and it is something of a puzzle that there is as yet no evidence that it has.[10]

Stars frequently have companions: two stars can be bound together by their gravitational attraction, orbiting around their center of gravity. For example, Sirius, the brightest star in the northern sky, has a faint partner which is in fact a white dwarf. Another much-studied star, a pulsar that is also a member of a binary system has provided a cornucopia of information relevant to the predictions of general relativity. We will have more to say about this binary pulsar in chapter 9.

On their way to death as white dwarfs, neutron stars, or black holes, stars can have a short episode in which their collapse is interrupted by a sudden explosion. This is a supernova explosion, a final surge of thermonuclear reactions ignited by the enormous rise in temperature as the star collapses under its own weight. When this happens, the star can throw off a large fraction of its content, the ashes of earlier reactions. These will include many elements more massive than the hydrogen from which the star was first formed, and in the intensely radioactive environment of the explosion still more transmutations take place. The gas cloud formed in this manner is what we see surrounding the site of a supernova like the Crab Nebula. The matter in the very early universe was almost all hydrogen and helium, with tiny traces of some light elements like lithium. *All* the heavier elements in the universe were made in stars and dispersed by supernova explosions—including the carbon, oxygen, and nitrogen which build our proteins. We are made from stardust.

There are several different kinds of supernova, triggered in rather different ways. For one type, the mechanism is analogous to that which leads to the flaring of a nova: material is drawn by gravity from a binary companion star. In this case a supernova is initiated when the accreted mass takes the star, which was previously a white dwarf, above the so-called Chandrasekhar mass limit. The Indian-born theoretical astronomer Subrahmanyan Chandrasekhar showed that if the mass of a white dwarf star were to exceed a critical mass (about 1.4 times that of the sun), it would inevitably undergo catastrophic collapse. As this happens, an enormous amount of energy is released, and this is what powers the resulting explosion, during which the supernova can outshine all the rest of the stars in its galaxy put together. All that is then left of the collapsed star is a black hole. The process, although quite complicated, is understood in sufficient detail to confirm what observations strongly suggest, that *all* of these supernovae have the same intrinsic luminosity, and that this luminosity is correlated with the directly observable way their brightness changes with time during the brief few days they shine so brilliantly. So if one can measure the way the brightness of such a supernova changes with time, one may deduce its intrinsic luminosity, how strong a source of radiation it actually is. And then from its apparent brightness its distance can be established: its apparent brightness is diminished by the square of its distance. So this class of supernovae provides ideal candidates as "standard candles" for the far-off galaxies in which they may occur. What makes them so important is that one cannot use Cepheid variable stars to establish distance unless they can be picked out as individual stars in a galaxy—and this is not possible for very remote galaxies. With this new, *direct* determination of distance, a comparison can be made with that inferred from the redshift of such galaxies. And when that is done, the results obtained by surveys of far-off galaxies published in 1998 suggest that the Hubble expansion of the universe is *speeding up*.

This suggestion is very surprising, since one might expect that the gravitational attraction between the galaxies would act to slow down the expansion initiated by the big bang. If instead it is speeding up, there must be some kind of negative pressure which pushes the galaxies apart, in contrast to gravity, which draws them together. But a force of this kind had been envisaged long ago by Einstein. Soon after he first formulated the general theory of relativity in 1915, Einstein turned his attention to cosmology. He sought a solution to his equations which would yield a cosmological model consistent with the Copernican cosmological principle—there should be no "special places" in his model universe. He also wanted his model universe to be *static* (i.e., there should be no special time either). There was no reason to suppose that on cosmological scales the universe was *not* static: Hubble was yet to come! Einstein thought that his equations of general relativity did not allow both these conditions to hold, and so modified the equations by introducing an extra term: the so-called cosmological term, or lambda term. But by 1931 he was convinced that there was no need for such a term, and in later editions of *The Meaning of Relativity* (Princeton, N.J.: Princeton University Press, 1950), the book he based on lectures he gave at Princeton in 1921, he added a footnote which reads:

> If Hubble's expansion had been discovered at the time of the creation of general relativity, the cosmologic member would never have been introduced. It seems now so much less justified to introduce such a member into the field equations, since its introduction looses its sole original justification—that of leading to a natural solution of the cosmologic problem.

But today there is a different justification, or at least a motivation, to consider such a term. This is because a small, but non-zero, cosmological constant (what Einstein called the "cosmological member" in the quotation above) can account for the acceleration of the Hubble expansion suggested by observations of the remotest galaxies. The supernova data that seem to call for something of the kind will no doubt be extended in the early years of the new century and are awaited with keen interest. However, supernovae in the distant galaxies may differ from those closer-by. If they do not have the same intrinsic brightness, and so cannot be regarded as standard candles, the whole chain of reasoning leading to an acceleration of the Hubble expansion is broken, and there may be no need for the cosmological term after all.

We also have good reason to believe that in addition to the matter we can see—the stars and obscuring clouds of gas—there must be a great deal more that is invisible, its presence only shown by its gravitational effects. This *dark matter* cannot all be "ordinary matter"—for example, dead stars which no longer shine—for that would have other, albeit indirect, consequences besides those due to gravity. So what is this mysterious dark matter? One possibility is that there is some new kind of elementary particle which interacts with ordinary matter so weakly as not yet to have been observed. And experiments are currently underway to seek out such WIMPs (weakly interacting massive particles), as they have been named. Theorists have come up with alternative possibilities, but until there is more direct evidence, the search for dark matter will remain one of the hot topics for research in astronomy.

There is one small, but important, exception to the Copernican cosmological principle of overall homogeneity one important feature of the universe which is almost, but not quite, uniform and homogeneous. This is the celebrated cosmic microwave background radiation. It was discovered in 1964 by Arno Penzias and Robert Wilson, who were working at Bell Telephone Laboratories, investigating a source of radio noise which might interfere with the satellite communications systems then being developed. They found that wherever they pointed their 20-foot horn antenna, the noise persisted, and after exhaustive effort eventually concluded that its source was not in our galaxy.[11] What they had discovered was the black-body radiation which perfuses the universe, an afterglow of the big bang in which it was created. As the universe expands, it cools. When the universe was some 300,000 years old, its temperature had fallen to around 3000 K.[12] For the first time it became possible for electrically neutral atoms to form, and the universe then became transparent to radiation.[13] The radiation that filled the universe then is what Penzias and Wilson discovered. This is what we now observe as the microwave background radiation. It was black-body radiation when the universe was 300,000 years old, corresponding to the temperature then, 3000 K. But now the universe is 1,000 times bigger, and it has cooled. The wavelengths of the radiation have been stretched by that same factor of 1,000, and the result is still black-body radiation, but now at a temperature of just 2.728 K (figure 3.6). Please note the precision to which it has been

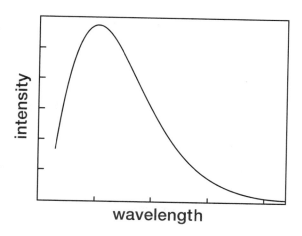

Fig. 3.6. The spectrum of the cosmic microwave background as measured by the COBE satellite. The data deviate from the shape of the curve Planck predicted for black-body radiation by less than the thickness of the line on the graph. It is remarkable that the cosmic microwave spectrum is the purest black-body spectrum ever observed. (From COBE datasets developed by the NASA Goddard Space Flight Center under the guidance of the COBE Science Working Group, provided by the NSSDC)

measured—about ten times better than can be achieved with a conventional mercury thermometer!

Satellites have carried sensitive detectors above the obscuring gases in the atmosphere to make impressively detailed measurements of this radiation. It conforms wonderfully to the precise form predicted by Planck's law; in fact it has the best black-body spectrum ever observed. And it is to a high degree uniform in all directions. But not entirely. One departure from uniformity, tiny but expected, is the Doppler shift produced by the combination of the motion of the earth round the sun, the sun through the Galaxy, the Galaxy within its local cluster, and even of the local cluster within its supercluster—which add up to around 600 kilometers a second! But beyond that there are tiny deviations from uniformity, at the level of about 1 part in 100,000 (plate 10). And that, together with the existence and size of the superclusters, turns out to be a vital clue to how all of the structures which fill the heavens, from superclusters to galaxies and the stars and planets within them, have developed from the big bang.

The first decades of the third millennium promise to be rich in new observations. For example, NASA's Microwave Anisotropy Probe (MAP) and the European Space Agency's Planck mission will tell us much more about the fluctuations in density in the very early universe, from which, in turn, we will learn more about how stars and galaxies were formed. Did stars come first, later to aggregate into galaxies, or was it the other way round, with matter first concentrating into great galactic clouds inside which stars were later to condense? We don't know, but we shall find out!

Multinational missions are launching the next generation of satellites to explore x-ray sources (such as NASA's Chandra[14] X-ray Observatory), which utilize greatly refined techniques to produce more sharply focused images than ever before. These will tell us about the emission of jets of matter the dense central regions of galaxies associated with the supermassive black holes, and about other high-energy regions of the universe, such as the hot gas remnants of exploded stars. The European Space Agency's international gamma-ray astrophysics laboratory INTEGRAL will be able to provide sharp images of gamma-ray sources and measure their intensity to high precision. Its launch is currently scheduled for October 2002, on a Russian "Proton" rocket. These and other such enterprises will allow us to resolve some of the challenging puzzles which twentieth-century astrophysics and cosmology passed on to the future.

4

CHANCE AND CERTAINTY

★ ★ ★

The Weird World of Quantum Mechanics

THERE IS LITTLE SCOPE FOR CHANCE IN THE DETERMINISTIC, classical world of Newtonian physics; any uncertainty would appear to be limited to the imprecision of our imperfect measurements. Although complex systems might defy analysis, we can try to deal with them by first considering the behavior of their constituent parts, following the reductionist strategy. The triumph of classical physics had its genesis in that approach. What could be a simpler system than a single particle moving under the influence of a known force? But even for so simple a situation, much subtlety and care was required before a general law emerge. The notion of *particle* needed to be refined, as did that of *force*. And a description of motion, of *kinematics*, had to be elaborated before even a start could be made. It was the clarity and precision with which Isaac Newton addressed these issues that set the agenda for the succeeding two centuries and more of classical physics.

It was a natural progression to go from a single particle to two, to many. From particles to extended bodies was a more difficult step, initiated by Newton himself. What about the forces, which produce the accelerations causing particles to deviate from motion in a straight line at constant speed (already recognized by Galileo Galilei as the "natural" motion of a body free from applied force)? Gravitation, for example: a force acting across empty space (what challenges were concealed here!) according to rules so simple and consistent that the twenty-four-year-old Isaac Newton could see that the *same* rules explained the fall of the apple from the tree, the monthly orbit of the moon around the earth, the rise and fall of the tides, and the orbits of the planets around the sun. These same simple laws of Newtonian dynamics and the universal attractive force of gravity are all that is needed to compute with precision the trajectories of spacecraft and artificial satellites, to arrange rendezvous in space or send explorers to the edge of the solar system and beyond.

There are of course other forces besides gravitation. Another pillar on which classical physics stands is electrodynamics, which may best be encapsulated in Maxwell's equations. Electric and magnetic forces, like gravity, act between spatially separated particles. In Maxwell's theory the forces are carried across the space between the particles by *fields*, physical quantities which extend through space. In Maxwell's electrodynamics they are not only fields of force (something already implicit in Newton's gravitational theory) but they have a dynamical behavior of their own. The electromagnetic fields, as well as the particles with which they interact, carry energy and momentum. Their behavior still conforms to the same basic dynamical laws of Newtonian mechanics, once one adjusts to the fact that fields have extension in space, in contrast to discrete, pointlike particles. These dynamical laws can be expressed through *equations of motion*. Newtonian mechanics and Maxwell's electrodynamics both describe systems that evolve in time according to a simple kind of equation. This means that if one knows enough about the system at some instant of time (the *initial conditions*), one can give a precise and unambiguous determination of the way it will behave in the future. The dynamics (for systems isolated from outside disturbance) is completely *deterministic*.

Pierre-Simon Laplace had speculated[1] that it might *in principle* be possible to predict the future from an exact knowledge of the present. But even in classical physics this was in practice at best an unrealistic dream. To follow the motion of the 10^{24} molecules of air in a bicycle pump is in any case not a sensible way to explain the fact that the air gets hot when the tire is inflated. The average and most likely behavior of a macroscopic system made up of a very large number of particles can be studied without a detailed description of the behavior of each of its microscopic or atomic constituents. One can describe the behavior of the air in the tire knowing only its pressure, temperature, and volume. This way of looking at things is the province of thermodynamics, a very practical science developed in its own right by Sadi Carnot and others concerned with designing better steam engines. But these same laws of thermodynamics can be derived from the application of Newtonian mechanics in a statistical fashion; for example, to the molecules in the air filling the tire. In that derivation, properties like temperature and entropy *emerge* and gain significance only when the number of particles in the system is very large. One may say that thermodynamics, statistical mechanics, and Newtonian dynamics exemplify a reductionist hierarchy within physics. It might well be possible to calculate from first principles the temperature at which water boils, using only the physics and chemistry of water molecules, but it would not be sensible to do so, nor would it add very much to our understanding of phase transitions (such as that between liquid and vapor), still less of the fundamental laws determining the forces between atoms.

Statistical mechanics and thermodynamics are important branches of physics in themselves; they also have an important role in the revolutions that transformed the deterministic view of the old classical order to bring about our present more fluid, and less secure, view of the universe. One of the triumphs of statistical physics was to show how the precise laws of thermodynamics for macroscopic systems emerge as consequences of the average and most probable behavior of their microscopic constituents. It was not necessary to propose the laws of thermodynamics as independent laws of nature; they can be deduced from the laws of mechanics. But statistical mechanics also exposed a profound, indeed fundamental, problem when applied to the study of radiation.

Imagine a foundry. Iron heated in a furnace gives off radiation; and the higher its temperature, the more intense is the radiation it emits. And as it gets hotter, the spectrum of the radiation (that is, the distribution of intensity over frequency, the relative intensities, let us say, of infrared to visible light and so forth) changes too. The frequency at which the maximum intensity occurs shifts, so that at say 600°C, the radiation is strongest in the infrared—you can feel the heat, but it is not yet "red-hot"—while at a temperature of 6,000°C, the temperature of the surface of the sun, the peak in the radiation spectrum is in the blue-green range of visible light, and there is copious radiation across the whole range of colors from red through violet, all the colors of the rainbow. One of the major problems confronting physics at the end of the nineteenth century was to explain what lay behind this familiar phenomenon, the shift in the peak in the spectrum which makes the iron glow first dull red, then brighter orange and yellow as it heats up.

It was already known from the laws of thermodynamics that the better a body is able to absorb radiation, the better it is able to emit; so an ideal emitter would also be an ideal absorber of radiation, for which reason it is called a *black body*. As we said in chapter 2, for a black body the spectrum of light radiated depends only on the temperature and not on any other properties of the body. This was the background to Larmor's prescient challenge: determine the black-body spectrum as a function of temperature.

Enormous progress was made in the last quarter of the nineteenth century, with advances both in the theory of radiation from hot bodies and (no less significant) in the precision with which new experimental techniques allowed measurements of the spectrum, especially in the infrared. It became apparent that thermodynamics on its own could not meet the challenge, but the deeper level of theory offered by statistical mechanics allowed additional input. The black-body spectral formula thus determined agreed tolerably well with the recent experiments in the infrared. But the formula had a shocking and unacceptable consequence: it predicted that, at any temperature, the *total* rate at which energy would be radiated was *infinite*. This "ultraviolet catastrophe," as it became known, was to undermine the whole edifice of classical physics. It brought down the deterministic, Newtonian laws of mechanics and the vision of a clockwork universe which, once set in motion, progressed in a predictable fashion, following rigid and ineluctable rules.

From Newton onward, classical mechanics had been developed and its formulation extended with great elegance and mathematical refinement. One of the quantities introduced in this progress is what is called *action*, a term first used by Pierre de Maupertuis in the eighteenth century in his study of mechanics. Action is a mathematical expression characterizing a dynamical process and is constructed from the product of the energy of the system and the duration of the process.[2] And it enters into the development of analytical mechanics, the formulation of Newtonian mechanics associated with the names of Joseph-Louis Lagrange, Carl Jacobi, and William Rowan Hamilton. At the dawn of the twentieth century, the resolution of the black-body challenge was to give a profound new significance to the concept of action by introducing a totally new fundamental constant into physics, the quantum of action, Planck's constant.

Max Planck had made a number of attempts before stumbling upon what turned out to be the correct approach to black-body radiation. It is often convenient in mathematics to approximate a continuum by a succession of discrete steps.[3] (The method of calculus is tantamount to making such an approximation and then regaining the exact result by allowing the discrete steps to become vanishingly small.) What Planck did was in effect to stop when the steps, though small, were still finite. He used what were by then standard methods, based on statistical mechanics, but supposed that the black body could emit or absorb radiation only in discrete amounts or *quanta*, each with an energy proportional to the frequency of the radiation. The factor of proportionality between energy and frequency is a new fundamental constant, denoted h by Planck (and so also by physicists ever since). This unit of action sets the scale of phenomena at which the classical mechanics of Newton no longer works.

If this were simply a mathematical device, a preliminary discretization to be followed by proceeding to the limit in which h becomes vanishingly small, the ultraviolet catastrophe would emerge once more. Classical physics, the physics of the continuum, would be recovered, with all its successes—and with its disastrous failures. But quantum physics, initiated by Planck in 1900, does not allow h to become vanishingly small. The Planck constant of action h has a magnitude that is indeed tiny when expressed in the everyday macroscopic units of engineering.[4] But as we shall see, for the microworld of atomic and subatomic physics it is by no means negligible. That it is finite, albeit small, averts the catastrophe of hot bodies radiating energy infinitely fast. And because it is finite, albeit small, there is an essential granularity to physics; the smooth regularity of the classical world has given way to the random, but statistically predictable, quantum jump. Our refined scientific instruments, and even our eyes, detect individual quanta. Quanta can be counted, one at a time; they can be emitted or absorbed, like particles.

Here then is a new dilemma. Newton, in his *Opticks*, had favored the possibility that light might be corpuscular in nature, that is to say, that light had particlelike properties. But since the beginning of the nineteenth century, and the study initiated by Thomas Young of the diffractibility of light and the other manifestations of wavelike behavior characterized by interference phenomena, the corpuscular theory had been abandoned. Most certainly, light *does* behave like waves, with peaks in amplitude reinforcing peaks, and peaks canceling troughs, just like intersecting ripples on a pool of water (figure 2.2). Maxwell's electrodynamics gives a beautiful description of light as a complex of electric and magnetic fields oscillating in space and time, quintessentially as waves propagating in empty space at a finite, universally constant speed. The speed of light in a vacuum, usually written as c, is another of the fundamental constants of nature. The historian Eric Hobsbawm allowed the start of the twentieth century to be postponed to 1914,[5] and if I may be allowed the same latitude, I would say that twentieth-century physics is dominated by h, recognized as small but finite, while nineteenth-century physics is dominated by c, recognized as large but finite. Since Maxwell, and the experiments of Young and Augustin Fresnel, of Fraunhofer and Michelson, no one doubted that light (indeed all electromagnetic radiation) was wavelike. To be sure, along with such wavelike quantities as frequency and wavelength, one could recognize and account for such mechanical properties as energy and momentum. But there was no place in this scheme for light as *particles*, as quanta, as *photons*, each carrying a discrete amount of energy and momentum.

The acceptance of photons as real physical objects, rather than merely as a mathematical device introduced to evade a theoretical problem, derives from the work of Einstein. In one of the four great papers he published in 1905, he gave an explanation of the *photoelectric effect*. (When light falls on a metal, electrons can be ejected, so leading to the flow of an electric current. It was for this work, not for his theory of relativity, that Einstein was awarded the Nobel Prize for physics in 1921.) This effect had been discovered by Heinrich Hertz and had been studied extensively by the experimentalist Philipp Lenard. Lenard was a leading member of a school of German experimentalists who were bitterly critical of what they perceived as a perverse dominance of mathematical theory over experiment and phenomenology. This was associated in particular with the English school of physics—and Lenard was an outspoken Anglophobe.[6]

There was a real tension between experimental physicists on the one hand, who continued to devise ever more ingenious apparatus and to discover thereby new and exciting phenomena, and a new breed of theoretically inclined physicists on the other, who never themselves got their hands dirty in the laboratory but used increasingly abstract mathematical formulations for their theories. This tension persists. Richard Feynman wrote: "There is a division of labor in physics: there are *theoretical* physicists who imagine, deduce, and guess at new laws, but do not experiment; and then there are *experimental* physicists who experiment, imagine, deduce, and guess." It is rare for a truly great physicist to make major contributions both to experiment and to theory; perhaps the last such was Enrico Fermi. The tension between the two camps can be corrosive, but it can also sometimes be creative, the source of a fruitful interaction.

The progress of quantum physics from Planck's work in 1900 to its flowering a quarter-century later in the invention of the quantum mechanics that was to displace classical Newtonian mechanics owes much to "abstract scientific men," men like Planck, Einstein, Niels Bohr, Werner Heisenberg, Erwin Schrödinger, and Paul Dirac (figure 4.1). Planck was never comfortable with the theory that sprang from the seed he planted,[7] nor was Schrödinger; and Einstein struggled vigorously to oppose some of its consequences, but in vain. Ultimately there could be no compromise: classical mechanics had to yield in its entirety to the new quantum mechanics of the 1920s. That is not to say that classical mechanics is entirely dispensable: it is now regarded as an approximation that emerges from quantum mechanics in situations where h

Fig. 4.1. Founding fathers of quantum physics: (a) Max Planck (© Emilio Segrè Visual Archives, The American Institute of Physics); (b) Albert Einstein; (c) Niels Bohr (© Emilio Segrè Visual Archives, The American Institute of Physics); (d) Werner Heisenberg (Niels Bohr Archive, Copenhagen); (e) Erwin Schrödinger (© Photo Pfaundler, Innsbruck); (f) Paul Dirac (© Emilio Segrè Visual Archives, The American Institute of Physics).

may safely be neglected. And so it is that in most of our day-to-day experience we are ignorant of quantum phenomena and the world seems to be ruled by Newton's laws.[8] However, we *do*, in fact, live in a quantum world. The mystery is why, for many purposes, it seems to obey the deterministic laws of the old order.

So what, then, is quantum mechanics? One way to answer this would be to attempt to give the rules by which quantum mechanics allows one to calculate the observable phenomena of atomic physics (for example, the wavelength of the light emitted by a sodium vapor lamp); or to explain how atoms combine to make molecules, or even such basic facts as the stability of atoms. However, that would require a very different book from this one—one that would have far more mathematics! Instead, let me give some idea of the radical differences between the basic concepts of quantum mechanics and those of classical mechanics.

In classical mechanics, at least in principle, if one knows the forces acting on the particles of a system, and the position and velocities of the particles at some initial instant of time (the "initial conditions"), one can precisely predict their subsequent motion—which is to say, that motion is completely determined. The quantum-mechanical description is quite different. Even the initial conditions cannot be specified in the same way. The difference springs from an aspect of quantum mechanics which has entered popular vocabulary, if not popular understanding: Heisenberg's uncertainty principle. As we shall see, it is not possible to measure the position of a particle without disturbing its velocity; neither can the velocity be measured without disturbing its position. From this it follows that there is no way to give a precise "specification of the position and velocities of the particles at some initial instant of time." Instead, the way to specify a state of the system is to give the values of as many quantities as *can* be measured without disturbing any of the others.

Heisenberg's uncertainty principle derives from the fact that the act of measurement will in general force a quantum system to change its state. This means that a subsequent measurement of some other property will not in general yield the same result as it would have done had the first measurement not been made. Now, in the world of classical mechanics, one could strive to make only a very *gentle* disturbance to the system in the first measurement, so as to produce only a negligible consequence for the second. But in our quantum world there is a conflict between the gentleness of the disturbance and the accuracy of the measurement.

To be specific, suppose that we desire to measure both the position of a particle and its momentum. (Momentum is what Newton had called the "quantity of motion"; it is simply the product of the mass and the velocity of the particle.) To determine the position of our particle, one might try to see where it is. To do this as precisely as possible, we would use short-wavelength light, for just the same reason that bats, which rely on echoes to locate and catch small insects, use high-pitched, sounds of short wavelength for their aural radar. In fact, whenever one is using waves to seek out the position of something in space, one finds that the longer the wavelength, the coarser the detail it can render. It's rather like the need to use a fine pencil to make a detailed drawing. So to make a precise measurement of the position of the particle, we will want to use of short wavelengths.

But there now arises a conflict between our wish to pinpoint the position of the particle and our wish to measure its momentum. In order to see the particle, we have to shine light on it, which in effect means to bombard it with a hail of photons. And as they bounce off the particle, they will disturb its momentum. To make this disturbance as small as possible, we might try to use a single photon. (We must have at least one.) Since "looking" means allowing the photon to interact with our particle—to "bounce off of" it— and since the photon behaves like a particle when it scatters, it will inevitably make our particle recoil, and thus change its momentum. So as to minimize this recoil—this disturbance to the momentum of the particle caused by our attempt to measure its position—we might try to use a photon with only a low momentum. According to quantum mechanics, that means that it is a photon corresponding to a *long* wavelength of light. To minimize the uncertainty in position, we must use short-wavelength light; to minimize the uncertainty in momentum, we must use long-wavelength light. That is what Heisenberg's uncertainty principle is all about.

What Heisenberg showed was that this uncertainty can be described mathematically. He showed that the uncertainty in the position multiplied by the uncertainty in the momentum could never be reduced below Planck's constant divided by 4π. This limitation does not just affect our knowledge of position and momentum. It is quite comprehensive, and it destroys completely the basis for Laplace's dream. Recall that in classical physics a knowledge of the present allows a determination of the future; given the initial conditions of a body, the classical equations of motion govern its evolution forever more. But the initial conditions always involve pairs of variables, like position and momentum! So Heisenberg's uncertainty principle means that we can never be sure about the initial conditions; or, to put it another way, quantum mechanics says that a particle does not have simultaneously a precise position and a precise momentum. We cannot even know the present; we most surely, therefore, cannot know the future.

But what quantum mechanics does allow us to determine are *probabilities*, and not just in some vague and fuzzy sense: one may calculate with precision and accuracy the probability of any specified outcome resulting from some given initial state of a well-defined system.[9] In some special circumstances, this probability becomes a certainty: as the mathematicians would say, the probability then has a value of 1; that is, the outcome in question has a 100% chance of taking place. But such circumstances are rare.

To determine the probability of any change in a quantum-mechanical system, one has first to calculate what is called the *amplitude* for the transition between the initial and the final state of the system. The probability is then the square of this amplitude. In just the same way, to understand the interference phenomena so characteristic of waves one calculates first an amplitude (the "size" of the wave), from which the intensity (the "strength" of the wave) is then derived by taking its square. In the amplitude, interference is manifested by different contributions to the resultant wave reinforcing one another here or canceling one another there—as said before, like ripples on the surface of a pool. The quantum-mechanical amplitude has exactly this wavelike property, which is why interference phenomena occur in quantum mechanics, even quantum mechanics applied to particles like electrons. It was predicted by theory, and confirmed by experiment, that electrons exhibit the characteristic interference and diffraction behavior of waves (figure 4.2). Indeed, the formulation of quantum mechanics that we owe to Schrödinger

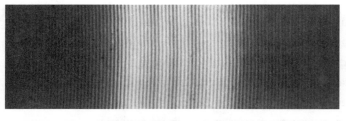

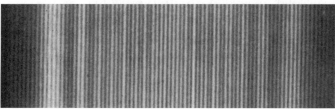

Fig. 4.2. Diffraction patterns. The pattern on top was made by light; that on the bottom by a beam of electrons. (From: J. Valasek, *Introduction to Theoretical and Experimental Optics* (John Wiley and Sons, New York, 1949); and H. Dücker, "Lichtstarke Interferenzen mit einem Biprisma für Elektronenwellen," *Zeitshrift für Naturforshung*, Vol. 10A (1958))

is called wave mechanics, and in Schrödinger's approach to quantum mechanics, the burden of the calculation is to solve his equation, which is in the form of a wave equation.[10] The amplitude is then constructed from the wave-function, that is to say from the solution to Schrödinger's equation.

In quantum mechanics one does not find pointlike particles following well-defined trajectories, as one does in Newtonian mechanics. The quantum-mechanical model of the atom is *not* a miniature solar system, with electrons orbiting the nucleus like planets around the sun—although Bohr's atomic model of 1913 was just like that, a curious chimera of classical orbits with superposed quantum rules. The state of motion of the electron is instead characterized by its wave-function, which is spread out in space. In the absence of external influence, some states, like the wave-functions which describe them, have a form which persists unchanged in time; these are the so-called stationary states of the system. An external disturbance may result in a transition between stationary states, and such a transition is what is termed a *quantum jump*. Knowledge of the nature of the external disturbance allows one to calculate, with precision, the probability that the jump will occur.

Although there is no dispute about how to *use* quantum mechanics, there is still a very lively and profound debate over its very foundations. Physicists still cannot agree on what quantum mechanics *means*. I will have to return to that later!

Initially formulated for the description of a single particle, the extension of these ideas to systems of many particles followed immediately. By then the distinction between particles and waves had become blurred. The electromagnetic field of Maxwell, when subjected to the rules of quantum mechanics, still yielded wavelike properties, such as diffraction and interference and so forth, but particlelike properties as well. The field is related to the quantum amplitude for the presence of *photons*, the quanta of light, discrete particles carrying each a definite energy and momentum. The duality between the particle- and wavelike modes of description is perhaps best summarized by saying that the field propagates as though it were a wave but interacts as though it were a particle. The quantum-mechanical treatment of the electromagnetic field is an example of *quantum field theory*. And it became apparent in the late 1920s that just as quantum mechanics applied to the electromagnetic field required a particlelike aspect of what was classically regarded as a wave, so also the phenomena classically regarded as particles—electrons, for example—must also be regarded as the quanta of fields.

If Maxwell's electrodynamics—the theory of electric and magnetic fields, charged particles, and their mutual interactions—was the paradigm for a classical relativistic theory, so now quantum electrodynamics—the theory of the electromagnetic field that has photons as its quanta interacting with another field that has electrons as *its* quanta—became the paradigm for a relativistic quantum theory. Particles, such as electrons, are themselves to be regarded as the quanta, the particlelike manifestations, of *fields*. The electron field provides the quantum amplitude for an electron in any given state of motion. And the electron field satisfies a wave equation.

A beam of light can be regarded as a shower of photons, so it is no surprise that photons can be created and destroyed. At the press of a light switch, a myriad of photons are created, streaming out of the lamp; when they are absorbed—for example, in the retina of your eye—they are destroyed. But it is not only photons which, as the quanta of fields, can be created or destroyed; the same is also true for electrons. A consequence of relativistic quantum field theory is that particles like electrons are not immutable and permanent. They too can be created or destroyed! The electron emitted from the nucleus of a beta-radioactive atom like ^{60}Co (cobalt-60, used as a source of radiation in medical applications) did not exist before the nucleus spontaneously decayed (another example of a quantum jump!) to form a nickel nucleus. It was created at that same instant of nuclear transformation.

For all the travails through which quantum field theory has passed, and they were many—and today even the field theory of pointlike particles is yielding to a theory of strings as the basis for understanding the structure of matter—the fundamental principles of quantum mechanics have survived all challenge and seem to be as secure as ever.

But we still do not quite know what they mean! The problem is centered on the nature of the quantum jump. Suppose, for simplicity, that we consider a system which has only two distinct basic states, any other state being constructed from these two. Certain real physical situations may be described in this way, for example, the polarization state of a photon. The polarization of light was studied by Thomas Young and is to some extent familiar to anyone who has used polarizing sunglasses or a polarizing filter on a camera. Light reflected from a pool of water is polarized, which means that it vibrates more strongly side to side than up and down. The two basic states of the polarization of a photon can be designated, for convenience, as ↔ (side-to-side) and ↕ (up-and-down). (Any other state of polarization is then a suitable combination of these two.) Suppose that I am wearing 100-percent-effective polarizing sunglasses; then the ↔ photons are blocked entirely, while the ↕ ones pass through unobstructed (figure 4.3). Now suppose that I tilt my head to one side, so that instead of blocking photons polarized in the horizontal direction, they block those polarized at 45° to the horizontal. What then happens to a ↔ photon? The quantum rules give a 50 percent probability that it will pass through, and a 50 percent probability that it will be blocked. This is in itself not so strange, since classical physics yields a similar result, and you may confirm it for yourself by looking at the reflected light from a pool of water through polarizing sunglasses. But think about the implications for a single photon. Before it met the tilted sunglasses, it was a ↔ photon; and any polarization state of a photon can be expressed as a suitable combination of ↔ and ↕. (There was nothing very special about our choice of these two basic states, and we might as well have chosen instead two perpendicular states inclined at 45° to the horizontal. Then our ↔ photon was just

Fig. 4.3. *Photons and sunglasses. Left:* Vertically polarized light (↕) passes freely through the lens of the sunglasses. *Middle:* Horizontally polarized light (↔) is blocked. *Right:* Horizontally polarized light is transmitted by tilted sunglasses, but with reduced intensity, and emerges polarized obliquely.

in a combination of these two states, with equal amplitudes for the two components, yielding the 50 percent probability of passing through the sunglasses as observed.) But now suppose that I put on *another* pair of sunglasses, oriented in exactly the same way. Remember that these are 100-percent-effective sunglasses, so that if a photon passes through the first, it will be unaffected by the second. This can only mean that if it had passed through the first pair of sunglasses, its state was such that it passed through the second with certainty, and that means that its state was no longer the state in which it started: it has *made a quantum jump* to the state which passes freely through the tilted sunglasses—call that ↗. We may describe what has happened by saying that in effect the tilted sunglasses measure the polarization by asking, Is the state ↗? And if the answer is yes, a subsequent measurement will again yield the same positive answer. The initial state was ↔, which was a superposition of ↗ and the other state corresponding to tilting your head in the opposite direction—call that ↖. After the first "measurement," if it passed through the sunglasses it was in the state ↗, and if not, it was in the state ↖. It is as though the act of measurement has forced the photon to choose. The wave-function of the photon before the measurement was a combination of ↗ and ↖; after the measurement, if it passed through the sunglasses, it was pure ↗. The wave-function has *collapsed*. This collapse of the wave-function occurs whenever a measurement is made. It is at the heart of the unresolved debate over the foundations of quantum mechanics, a debate which has gained impetus in recent years because of deep theoretical investigations by John Bell and new experimental techniques that have allowed ever more subtle tests. These tests confirm that indeed the rules of quantum mechanics are satisfied, and those of classical mechanics are not.

What seems to happen is that an undisturbed system evolves in time in the manner prescribed by Schrödinger's equation. This equation is itself no less deterministic than the equations of motion formulated by classical physics. Its solution, the wave-function, gives as complete a description of the system as is possible. But it does not in general determine the outcome of any measurement which might be performed on the system; it only ascribes precise probabilities to the possible outcomes. However, if a measurement is *immediately* repeated, the outcome is *certain* to be the same as on the first occasion. This can only mean that associated with the act of measurement there has been a *discontinuous* (or at any rate imperceptibly rapid) change of the wave-function. Somehow and somewhen during the process of measurement the wave-function appears to have abandoned its smooth, deterministic evolution according to Schrödinger's equation, and made instead an abrupt change.

This is puzzling enough in itself, even for the simple situation I have tried to describe. But it becomes even harder to understand if one asks such questions as, When does the measurement take place? Schrödinger provided a famous and provocative illustration of this puzzle—from which it would seem that he had no great love for cats! Even though the rules of quantum mechanics were first formulated for systems on an atomic scale, there is no reason why they should not also be applied to macroscopic systems too. Indeed, they must be so applied if we are to accept that quantum mechanics is of general and universal validity. (To date many experiments confirm its validity, and none cast doubt on it.) So let us consider a macroscopic system, the gross behavior of which is determined by some atomic process. After all, *every* measurement of an atomic system has to be of this kind, since our apparatus will be macroscopic and the count of a detector, or the movement of a pointer on a dial, is the sort of gross behavior of which I am thinking. The atomic process in Schrödinger's illustration is the radioactive decay of an atom, and we may suppose that the atom has a 50 percent probability of decaying in an hour. The decay of the atom is to be detected by a suitable apparatus, which therefore has a 50 percent chance of responding in the course of an hour. But now comes the devilry! The apparatus, atom and all, are to be sealed inside a box; and the apparatus is linked to a gadget which will release into the box a lethal dose of cyanide, and along with the radioactive atom, the apparatus, and the phial of cyanide, a cat is to be placed inside the box. The box is sealed up for an hour, at the end of which time there is a 50 percent probability that the atom has decayed, the apparatus has responded, the cyanide has been released, and poor kitty is dead. Equally, there is a 50 percent probability that kitty is alive and well (for out of kindness we have put a saucer of milk and other goodies in the box along with the cat). We now open the box, and find either a dead cat or a live cat. Schrödinger now invites us to ask, When does the measurement (cat dead/cat alive) take place? When we look? When the atom decays? Suppose we consider the wave-function describing the state of the atom: when does it collapse? Or suppose we consider the wave-function describing the whole content of the box, cat and all: when does *it* collapse? It is hard enough to imagine the state of the atom after an hour as being described by a wave-function which, until the act of measurement, is a superposition of two components, each with the same amplitude, one of which describes an undecayed atom while the other describes an atom which has decayed. But can we contemplate a wave-function describing such a superposition of dead cat and live cat? Are we really prepared to accept that until we open the box and "perform the measurement," there is some sort of superposition of dead cat and live cat inside the box, each occurring with equal amplitude after an hour? And if not, where should we make what John Bell called "the shifty split" between the atomic system following the laws of quantum mechanics and the classical cat? How are we to think of ourselves as quantum systems?

Of course no one has ever seen a quantum superposition of a live cat and a dead cat. But we *have* seen a quantum superposition of a $\updownarrow$ photon and a $\leftrightarrow$ one! And until a measurement has been performed to establish whether it was the one or the other, that *is* its quantum-mechanical state. Likewise, if I have a radioactive atom and consider its state after some time has elapsed, its state *is* in a superposition of undecayed atom and decayed atom, until a measurement is performed to establish whether it is the one or the other.

Opening Schrödinger's box is in essence performing just such a measurement. But what of the cat? The resolution to the paradox is still somewhat controversial, but I believe the following argument, which is widely accepted. The experimental setup in the devilish box establishes a correlation between undecayed atom and live cat and between decayed atom and dead cat: if the atom has not decayed, kitty is alive, but if the atom has decayed, alas, poor kitty is dead. The state of the atom and of the cat are said to be *entangled*. The weird consequences of quantum entanglement require the system to be left undisturbed, isolated from the environment (hence the box). Now the atom is pretty well isolated from its environment anyway—it is hard to influence the decay of a radioactive atom. But the cat, and indeed any macroscopic object, is at all times undergoing interactions with everything around it; for example, through collisions with molecules in the air. Though these interactions will not either kill a living cat or revive a dead one, they are able to change the quantum state of the cat. Hence, they are enough to destroy—and rapidly, too—the coherence necessary to talk meaningfully of a quantum superposition of states of a macroscopic object like a cat.[11] It is the rapid decoherence brought about by interactions with the environment that distinguishes between the microworld of the atom or the photon, where quantum superpositions are indeed observed, and the macroworld of the cat or the sunglasses, where superpositions are resolved. The only legacy of the entanglement of the states of the microworld atom and the macroworld cat is the correlation between the state of the atom and the state of the cat.

Einstein was never reconciled with the indeterminacy at the heart of quantum mechanics. His unease with the "Copenhagen interpretation" of quantum mechanics championed by Niels Bohr led to a series of exchanges in the years 1927–1931 between these two titans of twentieth-century physics, one of the most intense scientific debates of all time, conducted throughout with mutual respect. Searching for a weakness or inconsistency in the Copenhagen interpretation, Einstein proposed ever more ingenious "thought experiments" to demolish quantum indeterminacy. Bohr succeeded in finding flaws in each of them, culminating in a deft use of Einstein's own theory of general relativity to counter the most subtle of Einstein's proposals. Although Einstein yielded on this point, he continued to explore the deeper consequences of quantum mechanics, and in particular the conflict between the notion of *objective reality*, which would require a physical observable to have a value independent of the observation which measures it, and the fact that quantum mechanics can only ascribe probabilities to those values. Einstein speculated that perhaps quantum mechanics is *incomplete*; perhaps there are "hidden variables" which, if revealed, would restore classical determinism.

In 1935, together with Boris Podolsky and Nathan Rosen, he wrote a very influential and disturbing paper, which focused the debate on an apparent paradox. The authors showed that it was possible to devise a thought experiment—an experimental procedure realizable in principle, if not in practice—with which one could predict with certainty both the position and the momentum of a particle, even though Heisenberg's uncertainty relation dictates that precise knowledge of one of these precludes precise knowledge of the other. If position and momentum are elements of objective reality, a complete theory ought to be capable of describing the outcome of any experiment, but here was an example where quantum mechanics fails. So either quantum mechanics is incomplete—that is, there are hidden variables yet to be discovered—or the concept of "objective reality" needs to be radically reviewed. The thought experiment proposed by Einstein, Podolsky, and Rosen was simple in theory, but at the time beyond the scope of actual experiment. An alternative version of their experiment was devised by David Bohm, who proposed using the polarization of photons, rather than the position and momentum of particles, to illustrate the paradox, just as I have done in trying to explain the basic principles of quantum mechanics. He also devised an ingenious theory which incorporated hidden variables but was otherwise consistent with the predictions of quantum mechanics for the behavior of a single nonrelativistic particle. But still there was no experimental basis for answering the question, Do hidden variables exist?

A whole new approach to the issue was opened up in 1964 by John Bell, who gave a simple but profound analysis of the consequences of the existence of hidden variables. He considered the probabilities which give a measure of the correlations between the polarization states of pairs of photons as they might be measured in a simple experiment, similar to that described by

Bohm. But in his analysis he allowed the polarizing filters to be inclined at any angle with respect to one another. Quantum mechanics makes definite predictions for the correlations. Bell considered what might be the predictions of *any* theory with hidden variables that was consistent with special relativity. (To be consistent with the special theory of relativity, the setting of the apparatus to measure the polarization in one beam could not affect the result in the other beam, for that would imply an influence propagating faster than light.) He was then able to show that a certain combination of the probabilities could never be larger than 2—the so-called Bell inequality. But the prediction from quantum mechanics is that this combination *can* exceed 2, and indeed equals $2\sqrt{2}$ (about 2.8) for a suitable setting of the angle between the polarizing filters. So there is a sharp contradiction between the prediction of hidden-variable theories consistent with special relativity and the prediction of standard quantum mechanics. The question now became accessible to experiment: does Bell's inequality hold or not? The experiments have now been done by Alain Aspect and his co-workers in Paris[12] and Bell's inequality has itself been improved and generalized. The results have differed markedly from the predictions of hidden-variable theories, but have agreed perfectly with those of quantum mechanics.

The same advances in measurement techniques which allowed Aspect to test Bell's inequalities, together with the wonderful ability we now have to trap and manipulate individual atoms, have also shown the way to exciting prospects for future applications. Quantum entanglement like that between the photons in Aspect's experiments is the basis for a secure method for encrypting and communicating information, an application of great commercial interest. And the superposition of states (dead cat/live cat—or, more easily realized, decayed atom/intact atom), which is a fundamental feature of quantum mechanics, makes feasible a new kind of computer, a quantum computer.[13]

Another very lively topic for research which will surely continue into the next century is *quantum cosmology*: the application of quantum mechanics and its methods to the *whole universe*. For this purpose it is no longer possible to consider a separation between the system to be observed and the external observer. There is no external observer, supposedly so macroscopic as to satisfy the laws of classical physics, able to observe the subsystem to which we apply the rules of quantum mechanics. Nor do I find it easy to contemplate an *ensemble* of universes, all of them initially in the same state, to help make sense of a proposition which asserts that a given property has a determinate probability of being manifested. Nevertheless, this *is* an approach used by some quantum cosmologists. It seems to me that in *our* universe, a proposition is either true or false. Problems of this kind press against the limits of physical theory. They have gained urgency as experiment and observation point to surprisingly intimate connections between the physics of subatomic, subnuclear particles and the physics of the cosmos on the grandest scale. It seems that quantum effects in the early universe are the seeds from which galaxies have formed (more of this in chapter 11). Cosmology and cosmogenesis are as much dependent on quantum mechanics as is the physics of atoms—and chemistry, and the processes of life.

5

ORDER OUT OF CHAOS

★ ★ ★

The Emergence of Structure

LEONARDO DA VINCI MADE MANY WONDERFUL DRAWINGS OF turbulent water, of eddies and whirlpools, of ripples and waves (figure 5.1). These drawings reveal a fascination with the emergence of form and structure, of order out of chaos. Even in the most furious of ocean storms, there are patterns and recurrences. Like so much else in physics, the study of hydrodynamics, or water in motion, owes much to Newton. So for the moment let us set aside the uncertainties of quantum mechanics and return to the determinism of the classical laws of motion. Using them to predict the turmoil of a tempest, even the storm in a teacup when it is stirred, is a task of daunting complexity. To make a start, we might consider the motion of an individual droplet of water, pushed about by its neighbors. The motion of a nearby drop might be expected to be similar, so that if they start off close together, they will remain so, at least for some time. This is what happens in the gentle flow of water in a placid canal, for example. But if the flow is faster, or encounters obstacles, it becomes more irregular. As it becomes more

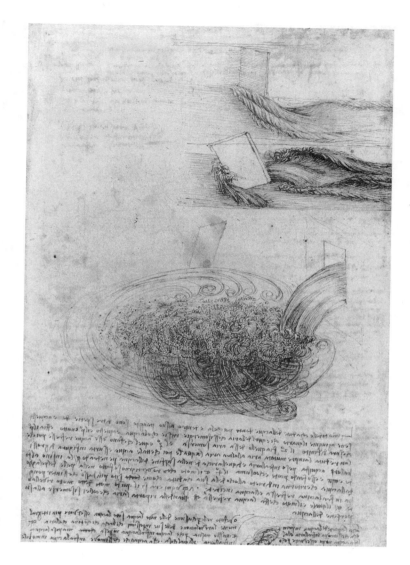

Fig. 5.1. A page from Leonardo da Vinci's notebook, with drawings of turbulent flowing water. (*Water* c.1507–9 from The Royal Collection © 2001, Her Majesty Queen Elizabeth II)

turbulent, drops which start out close together are rapidly separated from one another. The motion becomes *chaotic*. Turbulent hydrodynamic flows are chaotic—but they are not random (figure 5.2).

The planet Jupiter is the largest in the solar system. Through a telescope, its disc is marked by light and dark multicolored bands; these are clouds in the uppermost layers of its atmosphere of methane and ammonia, which can be seen to pass round the planet in about ten hours. These clouds are riven

Fig. 5.2. Examples of turbulent flow, showing regularities in the chaos. (Photographs of flows in soap films made by Maarten Rutgers (then at Ohio State University), Xiao-Lun Wu and Walter Goldburg (University of Pittsburgh))

by storms. Yet there is a giant feature which becomes visible in every rotation, and has persisted for at least a hundred years. This "Great Red Spot" is an anticyclonic system, a coherent structure arising from the tempests and turbulence, a wonderful example of the emergence of order and stability from chaos (plate 11).

Chaos does not occur only in complex phenomena like turbulence. It seems that for all the regular determinism of classical physics, chaotic behavior is the norm rather than the exception, once due account is taken of *nonlinearity*. For generations most of the study of mechanics was focused on linear systems or on linear approximations to more realistic situations. In linear systems, of which there are many in nature, small changes have small consequences. The ripples on a pond caused by a gentle breeze can be described by a linear theory, but the breaking of waves at the seashore cannot. It is only in the last twenty years or so that methods have been found to handle some of the mathematical difficulties which nonlinearity brings to a

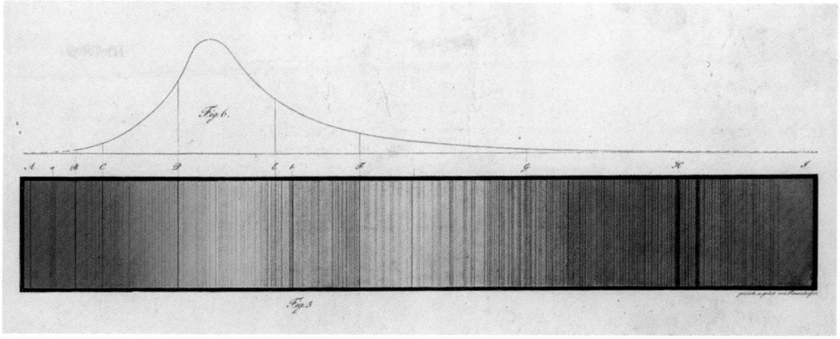

Plate 1. The dark lines in the spectrum of light from the sun, as seen by Fraunhofer. (With thanks to the Fraunhofer Gesellschaft, Munich)

Plate 2. The Milky Way as imaged by COBE, the Cosmic Background Explorer. (Photograph by E. L. Wright (UCLA), The COBE Project, DIRBE, NASA)

Plate 3. The Andromeda Nebula, very like our own galaxy and a member of our local cluster. (© Jason Ware)

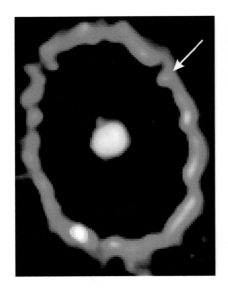 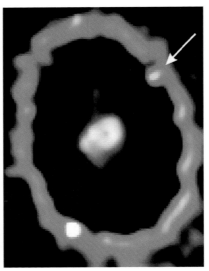

Plate 4. The supernova SN1987A. Hubble telescope observations show a brightening knot on the upper right side of a ring surrounding the supernova remnant. This is the site of a powerful collision between an outward moving blast wave and the innermost parts of the circumstellar ring. The collision heats the gas and has caused it to brighten. (Photograph by P. Garnavich of the Harvard-Smithsonian Center for Astrophysics, and NASA)

Plate 5. A patch of the sky that could be obscured by a pinhead held at arm's length. When magnified as in this picture, it reveals a dense field of remote galaxies, each containing tens of billions of stars. Photo taken in a ten-day-long observation above the Southern Hemisphere by the Hubble Space Telescope. (Photograph by R. Williams (STScI), HDF-S Team, NASA)

Plate 6. The Coma cluster of galaxies: thousands of galaxies each containing billions of stars. (Photograph by O. Lopez-Cruz and I. K. Shelton (U. Toronto), Kitt Peak National Observatory)

Plate 7. The host galaxy of the gamma-ray burster. The Hubble Space Telescope took this photograph some four months after the burster was first seen. (Photograph by K. Sahu, M. Livio, L. Petro, D. Macchetto, STScI, and NASA)

Plate 8. The Eagle Nebula. Within these great clouds of hydrogen gas, stars are forming, as the force of gravity concentrates the gas. (Photograph by Jeff Hester and Paul Scowen, Arizona State University, and NASA)

Plate 9. The Crab Nebula, the remnant of a supernova observed in 1054, is a gas cloud expanding around a pulsar. (Photograph by FORS Team, 8.2-meter VLT, ESO)

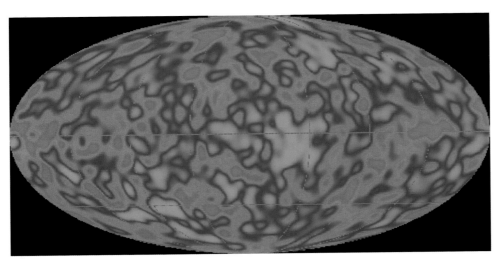

Plate 10. Temperature variations across the sky in the cosmic microwave background. They are represented by the different colors in this image and indicate structure present in the very early universe. (From COBE datasets developed by the NASA Goddard Space Flight Center under the guidance of the COBE Science Working Group and provided by the NSSDC)

Plate 11. Jupiter, with its bands of clouds and the Great Red Spot, an anticyclone which has persisted for at least a hundred years. (Photograph obtained by Voyager 1; JPL, NASA and the NSSDC)

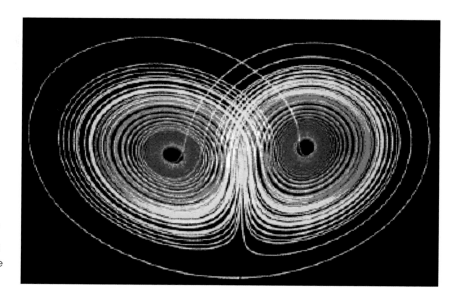

Plate 12. The trajectory of a chaotic system winding around the Lorenz strange attractor.

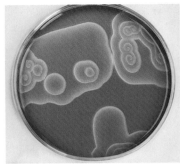

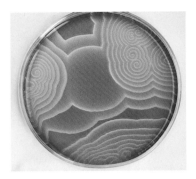

Plate 13. Spiral patterns spontaneously generated by the Belousov-Zhabotinsky reaction. (Photographs by Professor Peter Ruoff, Stavanger University College, Norway)

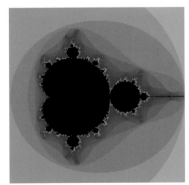

Plate 14. Computer-generated pictures of the Mandelbrot set. When tiny features of the boundary are magnified, they look very much like the larger picture from which they were taken.

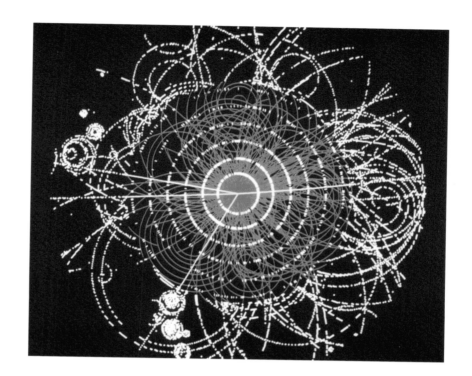

Plate 15. A simulation of the kind of event the ATLAS detector is designed to record. (CERN photos. © CERN Geneva)

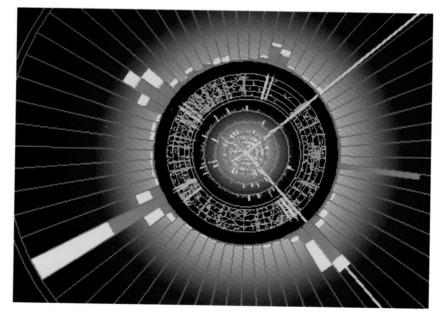

Plate 16. A pair of top quarks as imaged by the D0 experiment at the Fermi National Accelerator Laboratory, near Chicago. This end view shows the final decay products: two muons (turquoise), a neutrino (pink), and four jets of particles. (Courtesy of Fermi National Accelerator Laboratory)

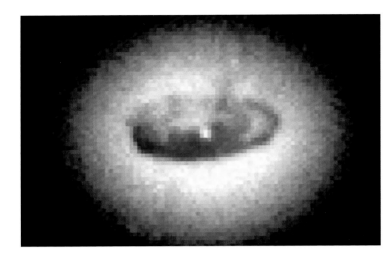

Plate 17. A spiral-shaped disk of dust is feeding a massive black hole in the galaxy NGC 4261. (Photograph by H. Ford and L. Ferrarese (JHU), NASA)

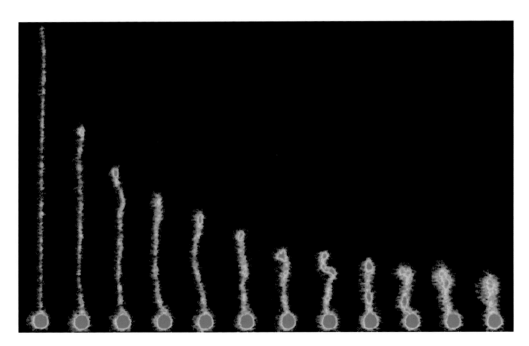

Plate 18. A single molecule of DNA, attached to a 1-micron polystyrene sphere, is stretched to its full length of 39 microns, and then allowed to relax. The successive frames in the photo are spaced at 4.5-second intervals. (Photograph by T. Perkins, S. Quake, D. Smith, and S. Chu. Excerpted with permission from Perkins et al., *Science* 264, 823 (1994). © 1994 American Association for the Advancement of Science)

realistic description of natural phenomena. In many cases remarkably coherent structures, like the Great Red Spot of Jupiter, emerge from the nonlinear interactions of very complex systems. And this notwithstanding the chaos in the fine structure of the motion, which makes it practically impossible to follow or predict its underlying dynamics in a deterministic manner. The coherent structures which emerge from this chaos of detail are themselves amenable to descriptive laws which have applicability across very different systems, as different as the atmosphere of Jupiter is from the spread of epidemics in populations. The "new science" of chaos addresses these structures and their laws.

What could be more regular than the motion of the planets around the sun, the moon around the earth? Yet it too can behave in a chaotic fashion. Imagine a simplified model of the solar system with just three bodies, call them Sun, Earth, and Moon. Set in motion under their mutual gravitational attraction they may for some time mimic the behavior of their actual counterparts in a familiar fashion, with the model earth in regular orbit around the model sun, and the moon in orbit around the earth. But over time this motion may change to one in which the earth and moon independently orbit the sun; then change back to the first kind of motion; and so intermittently go back and forth, first to one kind of motion, then the other. The whole sequence is completely deterministic: whether such alternations occur, and with what frequency, depends sensitively on the initial conditions. More alarming is the possibility that our model moon, or even our model earth, can in its motion escape completely from the vicinity of the sun. This could even happen in our own solar system! The stability of the system is so subtly dependent on the parameters of the motion, that although we may compute with a high degree of precision and certainty the motion of the planets in our own solar system over hundreds of millions of years—and it may seem that they none of them are about to escape—the uncertainties build up over time, and the prediction becomes insecure. We cannot be certain that the stability of the motions of the planets will endure indefinitely.[1] This startling conclusion concerning the stability of the solar system was reached by Jules Henri Poincaré in his submission for the prize offered by King Oscar II of Sweden in 1890; and it initiated the study of deterministic chaos. This is another aspect of what is called chaos; in the case of turbulent hydro-dynamics, the signature of chaos was the divergent trajectories of droplets which started close to one another, while here it is the divergent overall behavior of the system as a whole, which exhibits the extreme sensitivity to initial conditions that is the hallmark of chaos.

To better understand this sensitivity to initial conditions, it may help to consider an analogy. When a baker wishes to make croissants, he or she rolls out some dough into an oblong roughly one and three-quarters as long as it is wide (a mathematically pedantic baker would make it $\sqrt{3}$ times as long as it is wide). A knob of butter is then spread on the dough, which is then folded in three so as to obtain an oblong of the same shape as before (only smaller). The slab of dough is then turned through a right angle and rolled out until it is stretched to the same size and shape as at first; and then folded, turned, and rolled, again and again. The knob of butter is thereby blended in with the dough. More butter is added and rolled in. Each knob of butter has been stretched and squeezed, pulled this way and that. Its volume has not been changed, but it is now threaded in and out through the dough. The position of each speck can be followed from cycle to cycle, from one folding and rolling to the next. After each such cycle, its position can be calculated precisely: we assume, of course, that the baker is indeed a mathematician and has rolled the dough with impeccable accuracy. And so, cycle by cycle, the speck of butter appears now here, now there, its position shifted, or *transformed*, according to regular, predictable rules. The baker calls this mixing, and mathematicians call it mixing too. As it is not too difficult to see, a speck of butter in the exact center of the rectangle will still be there after each cycle. Such a point is called a *fixed point* of the transformation (there are two other fixed points of this transformation, one at the top left-hand corner of the rectangle and one diagonally opposite at the right-hand corner). Unless it is at one of the fixed points, each speck of butter gets moved about, threading its way throughout the dough; and two specks of butter which start off close to one another very soon become widely separated (figure 5.3).

The state of even a complicated dynamical system may be represented by a point (in what is called *phase space*), and the point moves as the system evolves in time. For a simple system, it may need only one variable to specify

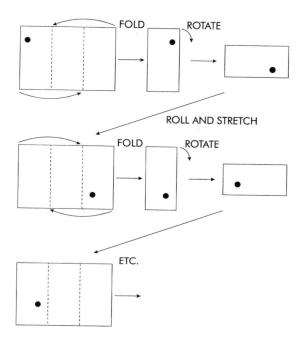

Fig. 5.3. The path of a knob of butter mixed into the dough of a croissant.

its configuration, and another (called its conjugate momentum) to specify how fast that configuration variable is changing. Then the phase space is two-dimensional and can be envisaged as a plane. The state of the system is then rather like the position of the speck of butter in the dough, and the motion of the system carries it about in the plane, in a way similar to the baker's manipulations carrying the speck of butter through the dough.

More complicated systems will need more variables than two for their phase space. Fortunately, mathematicians are not put out by the idea of multidimensional phase space—but they too, like ordinary mortals, may find difficulty in *picturing* it. One way to help capture the essential features of the phase space trajectories is to use a Poincaré section: from all the configuration variables and conjugate momenta a pair are selected which are to be emphasized. These define a plane which slices through phase space, and in general our trajectory describing the motion of the system to be studied will intersect this slice—again and again and again. The map of these points of intersection helps to provide a picture of the trajectory (figure 5.4). Another

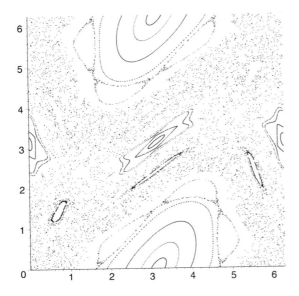

Fig. 5.4. The form of a three-dimensional curve can be illustrated by its intersections with a plane that slices through it. The points in this Poincaré section plot successive intersections with a plane of trajectories traced in phase space by a dynamical system. The continuous oval lines illustrate regular nonchaotic motions. But the scatter of points represents a single chaotic motion originating from slightly different starting conditions. (With thanks to Dr. Franco Vivaldi, Queen Mary, University of London)

picture (a *projection map*), emphasizing different features, is given by plotting a projection of the trajectory onto such a plane. It is rather like a radiologist giving the surgeon different views of an organ either by a tomographic section or by a direct photographic shadowgraph. Just as the baker's patterns of action provides a rule which tells where the speck of butter will move to from one fold and roll to the next, so the deterministic equations of motion for the dynamical system provides a rule that relates successive points on the Poincaré section of the trajectory.

The successive points on a Poincaré section are close analogues to the successive positions of the speck of butter in the baker's transformation. Again there can be fixed points of the Poincaré section, which correspond to a trajectory that always comes back to the same point every time it crosses the section. Or the trajectory might alternate between a pair of points on successive intersections, or hop between some larger number of points before coming back to the first intersection. In such cases the motion is *periodic* and will repeat over and over again, just like a pendulum swinging back and forth. The projection map of such a motion will be a closed curve. At its simplest it could be an ellipse; in a more complicated case, a figure eight; and so forth. A periodic motion is one which repeats; it may be a combination of two or more independent oscillations, like the rhythms of a drummer playing three against five. If the independent oscillations do *not* have frequencies related to one another by a ratio of integers, the motion is *quasi-periodic*. The periodic trajectories, and their quasi-periodic cousins wind their way around surfaces in phase space which are shaped like a torus (a torus is a ring-shaped surface, like a life buoy, or to return to the baker's shop, a bagel). For most dynamical systems there are other trajectories, besides these periodic or quasi-periodic ones, and they thread their way through phase space, passing infinitesimally close to the regular tori and to one another, a veritable tangle of motions. This is chaos, and it is not something rare and exotic, nor a property only of specially chosen systems. It is a general feature encountered with most dynamical systems for some broad domain of their parameters.

Up to now I have ignored another very general property of real dynamical systems: *friction*, or some similar mechanism which drains energy out of the system and into its surroundings. Because of friction a system left on its own will eventually settle down in some position of equilibrium: a swinging pendulum will eventually come to rest. But we can keep such a system going by feeding energy into it, driving its motion in the same way that the pendulum in a grandfather clock is kept swinging by the kicks of the escapement, driven by the falling weight. Again the motion may be described by a trajectory in phase space. We usually want to understand the way it behaves after it settles down from any transient effects brought about by "switching on" the driving mechanism which offsets the loss of energy to the environment through friction. Like the pendulum in the clock, it may settle down to a regular periodic motion, and it does not seem to matter very much how we start the pendulum off; the resultant motion is much the same. In this kind of motion, the trajectory homes in on an *attractor*, here that of a periodic orbit in phase space around which the trajectory cycles once every tick-tock, that is, once every second. One can have an attractor of this kind, with a period T, which remains qualitatively unaltered as some physical parameter is changed, until a further small change in that control parameter introduces an instability, a wobble which is manifested by a change from a regular beat at period T to one in which alternate beats differ in some regular fashion, and the system now approaches an attractor with a period close to $2T$. This is called a *bifurcation*, and further changes in the control parameter may lead to further bifurcations and further doublings of the period. For most dynamical systems, the sequence of such period doublings leads eventually to chaos. It was shown by Aleksandr Sharkovsky in 1964 that such bifurcations in continuous maps had unexpected general properties, and in 1978 Mitchell Feigenbaum found that the period-doubling cascade of transitions on the road to chaos exhibits quite remarkable *universal* features. The details of the system—which might be rolling flows in a convecting fluid or oscillations in an electric circuit—don't seem to matter. Identical numbers characterize for example, the ratio of the control parameter for successive period doublings. In other words, there are universal rules for the transition from order to chaos.

But what of chaos itself? The trajectory for a chaotic motion does not settle down to a limit cycle like those for periodic motions. It may loop around near some portion of phase space, looking deceptively like a periodic oscillation, with some feature repeating in a regular fashion, but then disconcertingly flip over to some other apparently periodic motion, with some quite different features. There is an executive toy which illustrates this very clearly. A pendulum free to swing in any direction hangs above a base-plate, in which is concealed an electromagnet. This magnet supplies an occasional kick which keeps the pendulum swinging, and three additional magnets pull the bob this way and that. The resultant motion is chaotic.

While a distracted businessman may amuse himself by trying to guess which way the pendulum will swing, a more serious challenge from the realm of chaos is to forecast changes in climate or weather. Here the system is so enormously complex, and there are so many dynamical variables needed to fully describe the temperature, wind velocity, cloud cover, rainfall, et cetera, over even a small region, that some sort of modeling, some kind of approximate description, is essential. In devising such a model it is important to capture the essential features. In 1963 Edward Norton Lorenz used an extremely simple set of just three equations to formulate a model of weather patterns that was able to mimic the dynamic flow of the trajectories in phase space, like those describing the real evolution of a weather system. He used equations first formulated by Barry Saltzman to describe the kind of fluid motion known as Bénard-Rayleigh convection, but was able to go further than Saltzman in exploring their numerical solution because he had access to far superior computing power. And he also had the genius to recognize the near universal application of what the solutions to the equations showed. As he expected, the trajectories traced out curves in his three-dimensional model of phase space. There were control parameters, numerical constants in the equations, which he could adjust from one run of computation to the next. He found stable solutions which settled down to limit cycles of regularly repeating behavior. But as the controls were changed, a new kind of attractor emerged. The trajectory traced out a curve which approached closer and closer to a remarkable two-lobed surface; looping around like a doodler, it shaded in one lobe, then flipped across to the other, then returned to the first, just like a fly erratically circling first one lampshade and then another.

But for all its apparent erratic behavior, the motion is not arbitrary: the equations are deterministic. Two solutions that start off in close proximity at some initial time and run close together for awhile begin to diverge from one another, so that after a short time it is impossible to infer from the behavior of one (say this solution shows the motion to be on lobe A) what is the behavior of the other (is it on lobe A or B?). This is, again, chaos. And the two-lobed attractor to which the trajectories approach is the famous *strange attractor* of Lorenz; it is believed that similar strange attractors may be features of the *real* day-to-day weather and the *real* longer-term climate (plate 12). This is why we can forecast the weather with reasonable accuracy on a time-scale of a few days, but beyond that our predictions become rapidly less and less reliable. Will it snow next January 1? The evolution of the pattern is so sensitive to the initial conditions or to subsequent disturbance that a butterfly flapping its wings might alter the forecast, blowing the weather from a trajectory that leads toward snow on New Year's Day to a trajectory that leads to no snow on that day. (Lorenz in fact coined the phrase "butterfly effect" to dramatize this extreme sensitivity to initial conditions shown by dynamic flows near a strange attractor.) So we can only hazard a guess, or calculate the odds.

Forecasting the weather is hard, because even if we could determine the equations governing the changes from one day to the next, they are subject to the chaotic sensitivity to the initial conditions: the details of today's weather. Still, we might have reasonable confidence that we understand the range of possibilities for the weather tomorrow, or next week, or even next year. But now suppose that the underlying equations themselves are altered, which would mean that the climate itself is subject to change. This can happen naturally, as indeed it did during the ice ages. It is possible for small changes, built up over many years, to lead quite suddenly to a dramatic change in the pattern of the weather, as the system flips over to a different trajectory altogether. Such a change may also be brought about inadvertently through human intervention, and that is why physicists are concerned about the causes of global warming, even though we cannot be sure of its consequences. The hints we have from the unusual behavior of the ocean current called El Niño or the alarming hole in the polar ozone layer may point to just such a shift in the global climate.

Strange attractors are an essentially universal feature in the phase space for real dynamical systems. Typically there will be small-amplitude, irregular outside disturbances or shocks, but nevertheless the system will show long-term stability in its overall behavior. Whatever its initial condition it rapidly settles down, approaching an attractor which governs the character of its motion. The attractor itself is *stable*, changing only a little when the system changes a little; it is the solutions, the individual trajectories, which exhibit the chaotic behavior we have been considering, not the overall patterns they weave in phase space. And it is not only physical systems that exhibit chaotic dynamical behavior. Similar mathematical models, similar equations, and similar time evolution with limit-cycle attractors and strange attractors have been used to describe the beating of a heart or the alternating balance between predators and prey in an ecological habitat. They have even been found to dominate the jittering cycles of the stock market.

Take a mixture of citric and sulphuric acid, potassium bromate, and an iron salt, and stir. If the concentrations are right, you will get a blue solution. Continue to stir, and it will suddenly turn red. Keep stirring, and it will turn blue again—and so on, regularly and dramatically alternating between red and blue. This is the famous Belousov-Zhabotinsky reaction, which behaves like a chaotic dynamical system. In fact, it *is* a dynamical system, obeying evolution equations not so different from those studied by Lorenz, and again reveal a strange attractor (one red lobe, one blue!). Under slightly different conditions, instead of a regular alternation in time, the red-blue oscillations become irregular; and still other concentrations of the reagents yield an alternation in space, organizing themselves into bands or spiral wave patterns of red and blue (plate 13). It has been argued by Ilya Prigogine that this ability of dynamical systems driven away from equilibrium to organize themselves into structures in space or in time is of wide generality and may account for some of the regularities observed in nature—and even in living organisms. The spiral pattern of the electrical waves which keep your heart beating obey equations similar to those of the Belousov-Zhabotinsky reaction.

For a more palatable example, stir a cup of coffee. In so doing, you supply a flow of energy at as large a scale as your cup allows. But as the viscosity of the coffee drains away the energy you put in, the coffee flows with smaller-scale eddies which eventually fade out to minute whorls and swirls too small to see.[2] You have generated turbulence, another face of chaos. In fully developed turbulence there seems to be a high degree of *self-similarity*, so that a magnified picture of a turbulent flow still shows its distinctive pattern of whorls and eddies and will continue to do so across changes of scale of many orders of magnitude. This property of self-similarity, when a magnified view of a system or structure has essentially the same form as the original, is characteristic of fractals, and fractals and their peculiar geometry are pervasive in the study of chaos. The word "fractal" was coined by Benoît Mandelbrot to describe such curves as coastlines or the outline of a cloud, which exhibit self-similarity across a wide range of scales—in an idealized, mathematical limit, across *all* scales. A very famous and beautiful illustration of a fractal curve is the outline of the Mandelbrot set often seen on psychedelic posters (plate 14). The strange attractor is a fractal, neither a line nor a surface, but somehow something in between. Behind these beautiful images lies more than just a mathematical curiosity. This new kind of geometry keeps cropping up in many branches of science, not just in the turbulent flows of wind and weather. Physicists are still not entirely comfortable with the need to deal with this kind of mathematical complexity. But I have no doubt that it will not go away, and that the challenge will have to be faced as we enter the twenty-first century.

Another unresolved question: what happens to chaotic dynamics if the classical, essentially Newtonian evolution underlying all the equations and behavior I have described in this chapter are replaced by the quantum evolution to which Newtonian dynamics is only an approximation? Trajectories in phase space will have to give way to something else, something more fuzzy, less deterministic. The study of this interplay between quantum uncertainty and classical chaos is still an active topic of research.

6

YOUR PLACE OR MINE

★ ★ ★

Relativity and Field Theory

FOR NEWTON, TIME AND THE PASSAGE OF TIME ARE UNIVERSAL and uniform: "Absolute, true, and mathematical time, of itself, and from its own nature, flows equably without relation to anything external, and by another name is called duration: relative, apparent, and common time, is some sensible and external (whether accurate or unequable) measure of duration by the means of motion, which is commonly used instead of true time; such as an hour, a day, a month, a year." (*Principia Mathematica*, edited by R. T. Crawford [Berkeley: University of California Press, 1939], p. 6.) The passage of time may be measured by astronomical ephemera, by the regular ticking of a clock, or the oscillation of a quartz crystal. If we synchronize our clocks, who could doubt that we will continue to agree on the time at which future events occur?

But it is not so; a consequence of the theory of relativity is that Newton's notion of absolute time had to be abandoned. Time is to be measured relative to an observer, and observers in different states of motion may measure different time intervals between the same two events, even if they are using identically prepared, accurate clocks. Indeed, the very notion of *simultaneity* must be reexamined.

Einstein started this reexamination in the 1905 paper in which he first presented the special theory of relativity: "When I say, for example, 'The train arrives here at 7,' that means that, 'the passage of the little hand of my watch at the place marked 7 and the arrival of the train are simultaneous events.'" (Leopold Infeld, who worked with Einstein in the 1930s, described this as "the simplest sentence I have ever encountered in a scientific paper.") On this we can agree, but that does not mean that the duration of the journey from Vienna to Berlin as observed by the stationmasters, and as observed by a passenger on the train, are the same. For the passenger, both departure and arrival times are determined by measurements in the train; but for the stationmasters, the departure time was measured in Vienna and the arrival in Berlin. Einstein's reasoning led him to a startling implication: moving clocks run slow! And the result is that the duration of the journey measured by the passenger is less than that measured by the stationmasters. To see why this must be so, imagine a kind of clock made from a pair of mirrors, one on the floor of the train carriage in which the passenger is riding, one on its ceiling, with a pulse of light reflected back and forth between the mirrors throughout the journey. A count of the number of reflections will give a measure of the time of the journey. But remember that light travels at the speed *c*, and that its speed is independent of the motion of the mirrors; this is a consequence of Maxwell's electrodynamics and is amply supported by experimental evidence. So to calculate the time of the journey, we need only take the total distance traversed by the light as it bounced back and forth between the mirrors, and divide by the speed it was traveling, namely *c*. For the passenger on the train, this distance is just the number of reflections times the distance between the mirrors (the height of the carriage). But for the stationmasters, the light does not go straight up and down, because the mirror on the ceiling will have moved closer to Berlin since the light was reflected from the mirror on the floor, and so between each reflection the distance will be *greater* than just the height of the carriage. The stationmasters will therefore draw the conclusion that the time taken for the journey was *longer* than that measured by the passenger. To them it would seem that the moving clock runs slow.

This "time-dilation" is one of the predictions made by Einstein in his special theory of relativity; but it is not just a matter of theory. The most accurate timekeepers we can construct at present are "atomic clocks," in which the oscillations of a quartz crystal (like the one in your wristwatch) are replaced by the vibrations of atoms of cesium; they are accurate to better than 1 part in 10^{13}. Two cesium clocks, once synchronized, will show the same time to an accuracy of one second in a million years. Yet if one clock is flown around the earth in a fast-flying plane, it will record a shorter duration for the flight than the one that stayed behind. This is a *prediction* of special relativity—but more than that, for the experiment has been done, and the prediction proves true. The special-relativity effect is tiny for the speeds we encounter in our everyday lives. But for speeds approaching the speed of light it becomes large, so that, for example, a muon (a radioactive particle constituent of cosmic rays) traversing the atmosphere at 99.9 percent of the speed of light (which is not unusual) will appear to live more than twenty times longer than one at rest.

By the same proportion, the distance traversed by the muon—that is, the thickness of the atmosphere—will be 30 kilometers as measured by an ob-server on the ground; but from the perspective of an observer moving with the muon, it would be not much more than 1 kilometer. This contraction was described by Hendrik Lorentz and George FitzGerald. The distance be-tween two events, and their separation in time, depend on the motion of the observer. If I am sent on a trip to the moon, so far as I am concerned the launch and the landing both occur at the same place; namely, *here*, where I am, in the spacecraft; whereas for mission control, they are separated by a quarter of a million miles. There's nothing mysterious, or even Einsteinian, about that—it's just perhaps an odd way of expressing things.[1] The duration of the flight is measured by mission control; I can signal to them when I arrive, and they can make due allowance for the time taken for my signal to reach them. And I can record the duration of the flight as measured by my on-board clock. As said before, I will record a duration *different* from that recorded by mission control. So we have two events, the launch and the landing, for which different observers determine different separations both in space and in time. Nevertheless, according to Einstein's special theory of relativity, there *is* something on which *all* observers will agree, and that is a separation in *spacetime*, calculated quite simply from the separation in space and the separation in time, using a formula reminiscent of Pythagoras's theorem.[2]

Einstein's special theory of relativity replaces the relativity of classical mechanics, which goes back to Galileo. Galileo recognized that only *relative* motions have physical significance.[3] There are no markers in space by which we can determine absolute motion, only other bodies with respect to which our motions become perceptible. As Newton expressed it:

> But because the parts of space cannot be seen, or distinguished from one another by our senses, therefore in their stead we use sensible measures of them. For from the positions and distances of things from any body considered as immovable, we define all places; and then with respect to such places, we estimate all motions, considering bodies as transferred from some of these places into others. And so, instead of absolute places and motions, we use relative ones; and that without any inconvenience in common affairs; but in philosophical disquisitions, we ought to abstract from our senses, and consider things themselves, distinct from what are only sensible measures of them. For it may be that there is no body really at rest, to which the places and motions of others may be referred.

Einstein went further and recognized that there are no markers in time either. Without either absolute space or absolute time, only *events* can be used as markers in spacetime, and the separation between events in space and in time depends on the motion of the observer.

What Galileo had appreciated was that the behavior of physical systems is unaffected by *uniform* motion (unaccelerated motion in a straight line). Different observers of the same phenomena may, of course, report different values for what they measure. Someone walking down the corridor of a train may have a speed of 105 kilometers per hour relative to an observer standing on a bridge over the track, but a leisurely 5 kilometers per hour relative to an observer buying a drink in the buffet car. Of course, there is a way to relate the two observations: the train is going at 100 kilometers an hour (and 100 + 5 = 105). So long as one is consistent in using the observations of either the observer on the bridge or the one in the buffet car, the physical behavior of what is observed follows from the same laws of motion. This is the principle of relativity. Newton's laws of motion are consistent with this principle, taken together with the simple addition law for velocities illustrated above. But Einstein saw the conflict between this and the fact that follows from Maxwell's equations, namely, that the speed of light, unlike the speed of the passenger walking down the corridor, of the train, is the same for the observer on the bridge or the observer in the buffet car traveling at 100 kilometers per hour relative to the bridge. He recognized that the only way to reconcile Maxwell's predictions with the principle of relativity was to change the rules that allow different observers to relate their observations one to another. The simple addition law for velocities had to go. And with it also Newton's mechanics—which does, however, survive as an excellent, approximate description of phenomena that only involve motions at speeds much slower than light.

To the passenger on the train, the cup of coffee he has bought in the buffet car sits at rest on the table in front of him; to the observer on the bridge, it is moving at 100 kilometers an hour. To the one it has zero momentum and zero kinetic energy; to the other both of these values are nonzero. But there are simple rules for relating what is seen by the passenger to what is seen by the watcher on the bridge, and similar rules for relating the values of all the kinematical observables measured by the two observers. These rules lie at the heart of special relativity. They are not the rules of Newtonian physics. What is *special* about special relativity is that it describes the way different observers moving relative to one another in a special way may relate one to another their measurements of the separation between events in space or in time, and indeed all the other kinematic quantities (those associated with motion, like velocity and energy and momentum) which are the domain of mechanics. The *special* motions are those between *unaccelerated* observers, just as in Galilean relativity. Galileo's hypothetical ship (see note 3) could not be pitching and rolling! In his *general* theory of relativity, Einstein relaxed the restriction to uniform relative motion of the observers, and the result was further to transform our conception of space and time, beyond even their fusion into spacetime as brought about by the special theory. Spacetime is no longer a passive arena in which the events of physics are played out. The geometry of spacetime is itself a dynamically active participant in the drama.

The change in kinematics introduced by special relativity already entails a change in Newtonian mechanics. This change was forced on Einstein by his determination to accept the consequence of Maxwell's electrodynamics, namely, that the speed of light in a vacuum is indeed a universal constant, independent of the motion of the source or the detector of the light. This fact, famously confirmed by experiments, both those culminating with Michelson and Morley's in 1887 and many thereafter, is not compatible with the transformation rules of Galilean relativity.[4] Something had to give; Einstein had the nerve to suggest that it had to be Galilean relativity, and the laws of Newtonian mechanics that incorporate it, rather than the electrodynamics of Maxwell. And yet, for all its daring originality, and for all the profound consequences it brings in its train, special relativity may still be considered as part of classical physics; it belongs to the "long nineteenth century" of physics.

But the marriage between special relativity and quantum mechanics—relativistic quantum field theory—definitely belongs to the physics of the twentieth century; and I do not doubt that the marriage between general relativity and quantum field theory will be accomplished in the twenty-first (more of this in a later chapter). This has been a marriage difficult in the extreme to negotiate, requiring some give and take on both sides. It would seem to require the bringing together of all physical phenomena into a single theory, including the unification of gravity with electromagnetism, of which Wolfgang Pauli had warned that men shall not join what God has torn asunder.[5]

To reconcile quantum mechanics with the special theory of relativity, one needs more than just a relativistic equation that generalizes Schrödinger's. Dirac had produced a relativistic equation for the electron in 1928, with wonderful consequences. It correctly described the electron's spin, how it behaves like a tiny gyroscope, and also like a tiny magnet. But more startling,

Dirac's equation predicted the existence of the *anti-electron* or *positron*, soon to be found experimentally. As a matter of fact, positrons had left their tracks on cloud-chamber photographs even before Dirac's paper, but had not been recognized for what they were. It was the formulation of relativistic quantum *field* theory, however, which opened the way to a new perspective on fundamental physics. For with relativistic quantum fields, particles are no longer immutable, permanent entities; they may be created and destroyed (as we have already seen for photons). Particles can be understood as the quanta of various fields, and the interaction of these fields allows us to understand how, for example, a neutron disappears, is annihilated, and in that same event are created a proton and an electron and an antineutrino. And it is just such an event that occurs when a cobalt-60 nucleus transmutes into a nickel nucleus, emitting the electron which the radiologist hopes will destroy a cancer cell. The experimental discoveries of subatomic and subnuclear physics in the second half of the century were all intertwined with relativistic quantum field theory. A bewildering variety of particles have been discovered, as experiments have explored higher and higher energies. Particle physics is high-energy physics, for to create and study the particles, most of which have only a fleeting existence before they decay, requires energy, and the more massive the particle, the more energy is required.[6] For as Einstein showed in another of his four great papers of 1905, $E = mc^2$, which is to say that energy and mass are interconvertible at a fixed rate of exchange given by the square of the speed of light.[7]

The theory that describes how the electromagnetic field interacts with electrons and positrons is a paradigmatic example of a relativistic quantum field theory. This quantum electrodynamics, QED, is astonishingly successful, but before its full predictive capability could be exploited, a very significant further development was needed. Most predictions of the theory are in the form of a systematic sequence of better and better approximations to the exact result, following a technique called perturbation theory.[8] That the theory was substantially sound was evident from the high level of agreement between experiment and prediction at the first level of the perturbation approximation; that the theory was deeply flawed was evident when the *next* level of approximation gave an infinite correction! At first, ways around this problem were found with seemingly ad hoc methods to "subtract" an infinite term, leaving a small finite part. This mathematical artifice did indeed enhance the agreement of theory with experimental results,[9] but it was clearly an unsatisfactory recourse and demanded a rigorous justification.

What followed was a new development in our concept of a "particle." An electron, for example, interacts with the electromagnetic field because of its charge, and so disturbs it. That disturbance in the field may be viewed as a fluctuating cloud of photons (the quanta of the electromagnetic field) surrounding the electron. The "bare electron" is "dressed" with photons—not photons free to propagate but "virtual" photons, emitted and then recaptured by the electron with a fleeting presence consistent with Heisenberg's uncertainty principle. The energy associated with this cloud is *infinite* and so contributes an infinite amount to the mass of the electron. But the observed mass of the electron is of course finite; what must be assumed is that the *bare* mass of the electron is also infinite, but such as to cancel the contribution of the photon cloud, leaving just that finite remainder which we observe as the mass of the physical, dressed electron. This cancellation of the infinities of perturbation theory was deduced in a systematic way by Richard Feynman, Julian Schwinger, and Sin-itiro Tomonaga. It is based on the elimination of the infinite difference between the bare and dressed mass together with a similar approach to the charge. The procedure is called *renormalization*, and the renormalization theory was one of the great achievements of midcentury theoretical physics. In later refinements it was shown how, for theories such as QED that are renormalizable (i.e., such that the infinities of perturbation theory can be removed in a consistent fashion), it is possible to rearrange the terms in the perturbation expansion so that only the finite, physical, "dressed" mass and charge appear, and the infinities are banished forever.

Not all interacting field theories are renormalizable. It seems miraculous that only those theories which pass the stringent test of renormalizability are needed to describe all that is observed in particle physics, and for the most part the renormalizability depends crucially on the fact that the interactions have a special kind of symmetry called a *gauge symmetry*. There is one glaring exception, however; gravitational interaction is *not* renormalizable! As it happens, for most of the purposes of high-energy physics, this problem may be set aside, because gravitational forces, although so dominant on the macroscopic scale, are quite negligible on the scale of particle physics. (the gravitational force between an electron and a proton, for example, is nearly 10^{40} times weaker than the electrostatic force.)[10,11] However, the challenge to provide a consistent quantum theory of gravity cannot be ignored, and as the twenty-first century dawns, it has become a prime focus for the efforts of many theoretical physicists. What makes this turn-of-the-century pursuit so exciting is that the goal may now be actually within reach. I will return to this in chapter 10 when I discuss *superstring theory*.

For the moment, let me emphasize that the marriage of special relativity with quantum mechanics has engendered a radical shift in the notion of what is meant by an elementary particle. It is no longer a pointlike, structureless mass, immutable, adamantine. Instead, it is a quantum excitation of a field, and although discrete and localizable, carrying the particlelike properties of mass, energy, and momentum, it also inherits the wavelike attributes of the field. The elementary particles having been replaced by excitations of fields, it is natural to ask, What then are the elementary fields? Since one cannot study a particle without taking into account its interactions with all the other species of particle/field, the answer is to some extent a matter of taste. What we call a proton, for instance, is sometimes and in some respects a neutron and a pion; and what we call a pion is sometimes a proton and an antineutron.[12] If big fleas have little fleas upon their backs to bite 'em; so protons contain protons, and so ad infinitum. One can even talk of a "hydrogen-atom field," but it is far more fruitful to think of the hydrogen atom as a bound state of a proton and an electron, or—if one insists on using the field-theory picture—as a composite excitation of the electron and proton fields. But even the proton is a composite that has, as more detailed probing reveals, a structure. And at the fundamental level, it is most useful to think of it as a bound state of yet more elementary particles called quarks. Following this line of argument has led to the prevailing "standard model" of high-energy physics. In this theory, the fundamental particles of matter are quarks and leptons (particles like the electron and the neutrino). But more of that in chapter 8!

7

MANY HISTORIES, MANY FUTURES

★ ★ ★

Feynman's Way to Do Quantum Mechanics

IN HIS 1965 BOOK *THE CHARACTER OF PHYSICAL LAW,*[1] RICHARD Feynman wrote, "I think I can safely say that nobody understands quantum mechanics." Although research on the foundations of quantum mechanics continues unabated, invigorated in recent years by the remarkable experiments made possible by advances in optics and the ability to trap and manipulate individual atoms, Feynman's assertion remains valid. But as Murray Gell-Mann said, "Quantum mechanics [is] that mysterious, confusing discipline, which none of us understands but which we know how to use. It works perfectly, as far as we can tell, in describing physical reality."[2] There are alternative ways to formulate quantum mechanics. The first of these, due to Heisenberg, used arrays of numbers which mathematicians call matrices. This was very soon followed by Schrödinger's wave mechanics (see chapter 4). Within a few months, these two apparently very different theories were shown to be equivalent. All this, and more, was accomplished in 1925–26. By 1930 the underlying mathematical structure that supports both the matrix and the wave formulations was set out in a textbook by Dirac.[3] In later editions of this book, there is a rather obscure passage, introduced with the warning, "This section may be omitted by the student who is not specially conversant with higher dynamics."

One of the ways that quantum mechanics is actually used exploits an approach due to Feynman, already outlined in 1942 in his Ph.D. thesis. This has its roots in the ideas in that obscure passage in Dirac's book. Feynman's approach gives identical results to those of Heisenberg's matrix mechanics or Schrödinger's wave mechanics—no surprise, since it does not differ in its foundations. But it comes into its own in the application of quantum mechanics to field theory. And it is hard to see how string theory would have been developed without it.

Classical mechanics is concerned with the evolution of dynamical systems in time. The instantaneous state of a system may be represented by a point in phase space, and the equations of motion then determine how that point moves, which is to say how the system evolves from an initial state to a final one. Quantum mechanics is also concerned with the evolution of dynamical systems from one state to another. But both the description of the states and the equations of motion are radically, conceptually different.

Suppose I wish to send a rocket to the moon. One way to use classical Newtonian mechanics to determine the trajectory of the rocket is to solve the equations of motion with appropriate initial data (position of launch and launch velocity), choosing them so as to reach the desired landing. Alternatively, I can set out to determine a "history" of the flight from launch to landing, a trajectory that starts from the launch site at some specified time and arrives at the lunar landing at a chosen time thereafter. And there is a way to do this. For any imaginable trajectory, any history, even one inconsistent with the laws of motion, one can calculate a quantity called the *action*, which depends on the position and time of the launch and the position and time of the landing, as well as on the path followed in connecting them (and on the forces acting on the rocket). A a wonderful and profound consequence of Newton's laws of motion is that the *actual* trajectory followed by the rocket, the one which *is* consistent with the laws of motion, is the one for which the action is at an extremum (either a minimum or a maximum).

This *action principle*, first suggested Maupertuis and later refined by Hamilton, is at the core of the "higher dynamics" referred to by Dirac. And it enters into Feynman's approach to quantum mechanics in a very remarkable way.

Before saying more about this, let me digress to point to another deep principle, this time drawn from classical geometrical optics. Pierre de Fermat had determined that the laws of refraction and reflection of light could all be subsumed into the proposition that the time taken for light to go from source to image along an actual light ray was a minimum, that is to say, less than for any imagined path differing from the ray itself.[4] Classical, geometric ray optics is an approximation to wave optics, without its characteristic phenomena of diffraction and interference. When the wavelength of light may be regarded as so small as to be neglected, ray optics emerges as a good approximation. Many before Feynman must have been struck by the analogy between Fermat's principle, which determines the paths of light rays as a limiting form of wave optics and the principle of least action, which determines the trajectory of a particle in classical dynamics. In optics the intensity of light may be calculated by taking the square of an amplitude (essentially the strength of the oscillating electromagnetic field—recall that light is nothing other than a kind of electromagnetic wave). This amplitude is itself the sum of contributions from many sources, or many paths, which can reinforce or cancel one another, leading to the phenomena of diffraction and interference characteristic of waves. In quantum mechanics there is also an *amplitude*. This too is to be calculated by adding up contributions that interfere with one another, reinforcing where they are in step and canceling where they are out of step. In Feynman's prescription for calculating quantum mechanical amplitudes, the analogy with wave optics is exploited to the full. But in quantum mechanics, by taking the square of the amplitude one calculates the *probability* of a quantum transition.

Feynman tells us to consider all imaginable motions that might carry the initial state at the initial time to the later state at the later time—just as is done in applying the action principle of Hamilton. For each such motion, each such imaginable history, determine the *classical* action. Divide this by h, the Planck constant of action, and identify the result with an angle. The quantum amplitude is then obtained by making a sum over all these histories

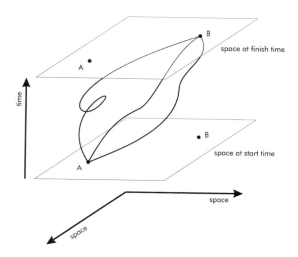

Fig. 7.1. Different "histories" for a particle going from A to B.

of a wavelike contribution from each, in a manner analogous to what is done in using Fermat's principle (figure 7.1). Now a wave has both a size and a *phase*, an angle which specifies where it is in its cycle of rise and fall. All histories, Feynman tells us, are to contribute with the same size; but the phase of each is just the angle we have calculated. So, says Feynman, make the sum over histories to determine the quantum amplitude, take the square to determine the probability, and there you have it! How wonderful, how bizarre! In some sense the evolution follows *all* of the imaginable histories, not just the one unique history singled out in classical mechanics. And just as in wave optics the waves reach their destination by following all possible paths, with the unique ray of Fermat's principle being nevertheless singled out in the short-wavelength limit, so also the unique classical trajectory may be recovered from Feynman's formulation by going to the "classical limit," in which Planck's constant is regarded as negligibly small.

What characterizes the quantum world is the notion of *interference* associated with all wavelike phenomena. Intersecting ripples on the surface of a pond can reinforce one another where they are *in phase*—that is, where they are moving up or down together—or weaken one another where one moves up and the other down. The same is true for the quantum-mechanical amplitudes calculated by Feynman's rules or by Schrödinger's equation. In this respect quantum waves are like ripples on a pond or the waves in the classical description of light. And just as with water waves, they can exhibit *coherence* (i.e., stay in step) for long periods of time or over long distances. There used to be a pair of pianists, Rawicz and Landauer, whose appeal was their ability to perform with temporal coherence. They would start a piece together, seated in adjacent rooms; the door between them was then closed until near the end of the piece, when it was opened, to reveal that they were still in time with one another. Likewise, a long column of guardsmen all in step, from those at the front to those at the back, shows spatial coherence. Interference, whether in classical optics or in waves on the surface of a pond or in quantum phenomena, only makes its effects manifest if there is a sufficient degree of coherence.

We see an especially vivid application of Feynman's "sum-over-histories" approach in quantum electrodynamics (QED) and in other field theories encountered in high-energy particle physics. It is also useful in the physics of solids, where, for example, *phonons*, the quanta of sound waves have a description very much like that of photons, the quanta of light. We have already seen that the marriage between quantum mechanics and relativity theory led to the recognition that particles—electrons, protons, and so on, as well as photons—may be created or destroyed, and quantum field theory is the best way to describe how this happens. Quantum fields can be regarded as acting to create or annihilate their quanta. When Feynman applied his prescription to quantum field theory, he was led to a very direct representation of how, through their interaction, the fields are responsible for such phenomena as occur in the scattering of particles or the more complicated processes studied at particle accelerators. When two particles collide, many different final states might emerge. For each, a probability can be calculated, and compared with what the experiments actually measure. The amplitude for any such transition—from the initial state of two colliding particles to any one of the possible final states—may be calculated according to the sum-over-histories prescription, and this then gives a representation of the amplitude in terms of what are called *Feynman diagrams* (figure 7.2). For example, if in the initial state an electron and a proton are present, and in the final state a proton, a positron, and two electrons (and this is possible!), one set of histories will have the initial electron interacting with the electromagnetic

Fig. 7.2. Richard Feynman at the blackboard. (© Emilio Segrè Visual Archives, The American Institute of Physics)

field so as to create a photon which is then absorbed by the proton, which in turn propagates for awhile before generating another photon, which then turns into an electron-positron pair. All of this can be summarized as a simple diagram (figure 7.3) showing the successive points at which interactions occur, creating or destroying particles, with the particles propagating in the intervals between these interactions. To each of these diagrams, then, there will correspond an aggregated contribution to the overall amplitude for the process, which may be calculated directly from some very simple rules. So the procedure is, or at least sounds, very simple. First, draw all possible Feynman diagrams contributing to the process. Then follow the recipe for calculating the contribution of each one to the overall amplitude. Finally, add them up, and take the square to determine the probability for the process (in effect, the rate at which it occurs).

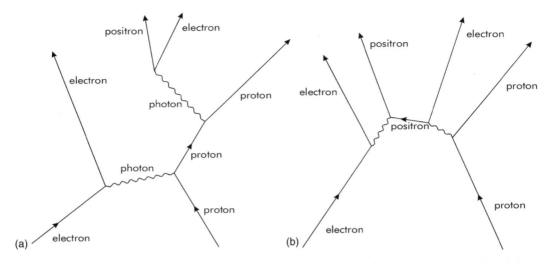

Fig. 7.3. (a) A Feynman diagram contributing to the process in which an electron and a proton become a proton, a positron, and two electrons. (b) Another Feynman diagram contributing to the same reaction.

Well, it's not quite as easy as that! For one thing, there are an infinite number of diagrams that contribute to any process, and it is always a tricky business to sum an infinite series of terms. In the case of QED it happens that the diagrams describing more complicated contributions carry higher and higher powers of a constant (essentially, the square of the charge on the electron) that determines the strength of electromagnetic interactions. This constant is the famous "fine structure constant," introduced by Arnold Sommerfeld in his very early attempt to incorporate relativistic corrections into the Bohr model of the atom. Numerically it is very close to $1/137$,[5] small enough so that one might argue that we need only calculate the simplest diagrams, say those which carry powers up to the fifth or sixth in this small number. After all $(1/137)^5$ is approximately 2×10^{-11}, which is very small indeed, but the precision to which experiments have been performed requires this kind of accuracy for comparison with the theory. This approach, based on successively higher powers of the fine structure constant, is *perturbation theory*; it is reminiscent of the way that perturbative effects of one planet's motion on that of the others may be calculated in classical celestial mechanics. As was said in the previous chapter, there were still problems to be overcome, the infinities which plagued the early approaches to the calculations. But the renormalization procedures of Feynman, Schwinger, and Tomonaga drew together and made secure the earlier subtraction methods and gave a general prescription for taming these infinities in QED. This approach works also for a restricted class of theories called gauge theories. By good fortune those needed for the microcosm to be described in the next chapter are gauge theories. The resulting formulation, at least for QED, where the coupling constant (the fine structure constant) is gratifyingly small, is truly impressive for its empirical success. The experiments are extraordinarily precise, and the theory agrees with them with extraordinary precision—to better than one part in a hundred million.

Subsequent theoretical refinements have recast the renormalization program so as to allow a formulation in which the infinities are absent from the outset. We now have a class of renormalizable relativistic quantum field theories which give finite results at every order in the perturbation approximation. And as said before, perturbative QED works amazingly well—and at the same time is surely a flawed theory, because it contains within itself the seeds of its own destruction! It can be proved that the perturbation expansion does *not* converge. In all likelihood it is what mathematicians call an asymptotic expansion, which means that taking into account terms of successively higher and higher order gives better and better approximations at first, but after a certain point the approximation begins to worsen and the results diverge. Fortunately for QED, the approximation is not expected to deteriorate until the first 137 or so powers of the fine structure constant have been incorporated. And $(1/137)^{137}$ is *truly* a negligibly small number!

Nevertheless there are nonperturbative features of quantum field theory which are the subject of research now and doubtless will remain so into the next century. These include *solitons*, which have particlelike properties—energy and momentum, for example—but are spread out in space; they have rather inelegantly been called "lumps" of energy and momentum. Solitons have a venerable history in the classical theory of waves, but it was only with the advent of electronic computers that their properties have been better appreciated. They play an important role in understanding the persistent structures which can emerge in chaotic systems. And they also figure prominently in the extension of superstring theory known as M-theory. We will have more to say on that in chapter 10.

One of the mathematical "tricks" used in the computation of Feynman amplitudes involves making the time variable a pure imaginary quantity.[6] This is an idea which goes back to Hermann Minkowski, who taught mathematics to Einstein and whose ideas on the unification of space and time deeply influenced Einstein's theories both of special and general relativity.[7] When the time variable is turned into an imaginary quantity, the phase expression in the sum-over-histories becomes what mathematicians call a decaying exponential, and the sum itself becomes a well-defined mathematical expression. It is in fact strictly analogous to the sums that occur in *statistical mechanics*. Here one is concerned with calculating correlations between physical quantities at separated points in a medium in thermal equilibrium. Corresponding to the quantum fluctuations that may be calculated from the Feynman expression are the *thermal* fluctuations in the statistical-mechanical analogue.

Founded in the nineteenth century by Boltzmann and Josiah Gibbs, statistical mechanics shows how macroscopic concepts, such as pressure, temperature, and entropy, can be related to the behavior of the myriads of atoms and molecules that make up any macroscopic quantity of matter (in a glass of water, something like 10^{25} molecules). Gibbs in particular derived the laws of thermodynamics from a statistical-mechanical foundation. Boltzmann was intrigued by the fact that although the laws of Newtonian physics do not distinguish an "arrow of time" (so that if we could instantaneously reverse the velocities of all the particles in a Newtonian system, the resulting motion would simply retrace the previous history like a film run backward), macroscopic systems nevertheless exhibit a clear distinction between future and past behavior. Ice in a glass of warm lemonade melts; friction brings a swinging pendulum to rest. Boltzmann's reconciliation of the microscopic time-symmetry of the equations of motion with the observed approach of macroscopic systems to a state of equilibrium introduced the idea of *probability* of state into classical physics. Left to itself a system will always evolve toward states of increasing probability and will only move toward less probable states

Fig. 7.4. Boltzmann's tombstone in the Central Cemetery, Vienna. (Courtesy of Austrian Central Library for Physics)

for episodes of very small duration. He linked this idea with a new understanding of the thermodynamic quantity called *entropy*, now related to the probability of the state.[8] These new ideas were subjected to harsh criticism, especially in the light of the theorem of Poincaré, which showed that a system left to itself will eventually—albeit perhaps after an extremely long time—return arbitrarily to a state close to its initial one. The "Poincaré recurrence time" is extravagantly long; for any realistic system, typically much longer than the age of the universe. In response to the objection that Poincaré recurrence undermined his reconciliation of macroscopic time asymmetry with microscopic time symmetry, Boltzmann is reported to have said, "You should live that long" (figure 7.4).

Statistical mechanics allows much more than just a derivation of the laws of thermodynamics. In particular it allows a study of the thermal fluctuations about equilibrium, which are of special importance in such phenomena as melting or boiling, what are called *phase transitions*. The relevant behavior of a macroscopic system follows from that of its individual atomic constituents by an averaging procedure that involves taking a "sum-over-configurations" of the system which, although microscopically distinct, are macroscopically the same. (For example, if all the molecules of the air in the room where I am writing this were to make a random change in direction, there would be a new configuration, microscopically distinct from the one before but of no significant difference on the macroscopic scale, and I surely would not notice it.) There is a very deep analogy between the Gibbs-Boltzmann sum-over-configurations and the Feynman sum-over-histories. If the amplitude to be considered in the quantum theory is for the probability that the state of a system is found to be unaltered after some finite time T (so that the final state is precisely the same as the original state, never mind what may have happened in the intervening time), the expression obtained when the time variable is made pure imaginary corresponds exactly to what is obtained in the statistical-mechanical calculation for a system at a finite temperature inversely proportional to T. And this lies behind the remarkable thermodynamic properties of black holes in general relativity (see chapter 10).

Mathematical analogies must be greeted with a fair measure of skepticism before being allowed to lead to inferences in physics. But this analogy between thermal fluctuations in classical statistical mechanics and quantum fluctuations in the Feynman formalism does deserve attention, and not only for black-hole thermodynamics. It leads to very interesting insights into string theory, where we encounter results mathematically equivalent to the statistical mechanics of certain systems at what are called *critical points*. At a critical point the fluctuations (thermal or quantum) have just the same kind of self-similarity we have encountered in the phenomenon of turbulence: a magnified version of the system at a critical point looks much the same as the original. For example, away from the critical point water boiling at 100°C has a different density from water vapor at the same temperature; there is a sharp boundary between the two. But at the critical point this distinction becomes blurred; inside every drop of liquid there are bubbles of vapor, and inside every bubble there are droplets of liquid—with the scale-independence characteristic of a fractal geometry. In the modern approach to renormalization something very similar to this is exploited. The field theory under study is supposed to be subjected to successive changes of scale. The consequential change in the resultant amplitudes is then calculated in terms of the "effective coupling constants" that determine the strength of interaction between the fields; these constants have to be adjusted at each scale change to maintain consistency. In some cases, important in both statistical physics and in field and string theory, there are fixed points for the effective constants, and it is these that determine the overall physical behavior.

In a classical deterministic system, the whole history is uniquely defined from the conditions at any one time. What statistical mechanics recognizes is that for macroscopic systems, which require the specification of an enormous number of variables to give a complete description of the state, it is futile and unnecessary to try to trace out in detail the ensuing motion. It suffices to make a coarser description, seeking only to determine the behavior on *average* of an ensemble of systems so similar to one another that they all share the same coarsened description. So, for example, to study the way that air escapes from a punctured tire, it is foolish to try to track the motion of every molecule involved. Statistical methods will suffice. But I wish now to turn to another aspect of the notion of histories in the evolution of a dynamical system.

Feynman's approach to quantum mechanics derives the probability for an initial state to evolve into any possible final state by making a sum over all the possible histories that might describe such an evolution. And the result is the same as that which follows from Schrödinger's equation. But that still leaves a nagging question: what is it that determines *which* of the possible outcomes in fact is actualized? What determines whether the radioactive atom has decayed in kitty's hour-long incarceration? How does quantum mechanics explain the "collapse of the wave-function," the quantum jump that resolves a superposition of possible outcomes into just one? Somehow the very act of measurement forces the system to decide, and somehow this is associated with the interaction between the observed system and the observing apparatus. The clue lies in recognizing that the measuring apparatus is macroscopic. Although it is still subject to the laws of quantum mechanics, its state involves so many degrees of freedom that, just as in statistical mechanics, a coarse-grained description becomes not only useful but in a practical sense mandatory. To answer the question, Is the cat alive or dead? requires taking an average of all the possible states consistent with the one outcome or the other. And in that averaging, the coherence that a detailed microscopic description would provide is lost, and with it the paradoxical superposition of live cat/dead cat. Schrödinger's equation describes how a quantum system evolves when it is left undisturbed. The unavoidable jostling by the environment, the continual battering by thermal agitation, which any macroscopic object undergoes, would need to be taken into account if we wish to consider the atom-plus-cat system as a quantum system, and that is futile. So we have to recognize that the fine details of the state of the cat are neither accessible, nor interesting, and so average over them. The result is to collapse the superposition of live and dead cat. All that remains is the correlation between the state of the atom (decayed/undecayed) and the state of the cat (dead/live).

This loss of coherence does not happen instantaneously; it takes a small but finite time, which can be estimated. For a cat decoherence takes place so rapidly as to be impossible to measure; but there are mesoscopic systems large enough to show this kind of decoherence, yet small enough to make the time for decoherence measurable. Experiments of great ingenuity—for example, those at the Laboratoire Kastler Brossel in Paris—have shown that this is indeed what happens, and thus one can say that we have seen the classical picture of the world emerge from the underlying quantum reality. In these experiments, the counterpart of Schrödinger's cat is the electromagnetic field in a tiny microwave cavity, a silica sphere only a tenth of a millimeter across. The "cat" is prepared by passing through the cavity an atom which is in a state of coherent quantum superposition between two distinct states—the analogues of Schrödinger's decayed/undecayed atom. So just as the cat is put into a state of superposition between live and dead, the field is put into a quantum state which is a superposition of two distinct states. To see whether the cat is alive or dead, we can open the box—but in the Paris experiments, a "mouse" is used to find out if the "cat" is alive or dead; that is, the state of the field is tested by passing another atom through the cavity. The results fully confirm the expected decoherence effects of the environment. Most physicists now accept that it is the rapid decoherence of the quantum state for any complex system which is not well isolated from its environment that resolves the paradox of Schrödinger's cat, and, more generally, allows us to understand how classical behavior, with no appearance of quantum interference, can emerge from the underlying quantum mechanics.

It is as though every measurement of a quantum system forces it to choose one of its own possible outcomes. If we open the box and the cat is alive, we know the atom has not decayed; if it is dead, the atom has decayed. But until we look, the outcome is not known. Some physicists would argue that prior to our moment of looking, the system is still in a coherent superposition of the two possible states (the cat, in effect, both alive and dead). A more plausible view is that the ceaseless interaction with the environment has already destroyed the coherence and even before we look to see what has happened, it has rendered the two possible outcomes incoherent with one another and so incapable of showing further interference (the cat is definitely one or the other—alive or dead). It is as though every interaction with the environment has had an effect similar to that created by a measuring apparatus: it has forced the quantum state to "make a choice." Elaborating on this line of thinking, one interpretation of quantum mechanics (one of the strangest) argues that every interaction is indeed of this kind, and that after the interaction all the possible outcomes are still present, still capable of interference with one another. But in *our* world, only the states consistent with the observations we have made are accessible to our future observation, although in *other* worlds any of the alternatives may be found. This *many-worlds interpretation* of quantum mechanics has its followers,[9] but I find it hard to accept. Perhaps I have done it less justice than it deserves!

8

MICROCOSM

★ ★ ★

The Standard Model of Particle Physics

IN 1939 PARTICLE PHYSICS WAS SCARCELY DISTINGUISHABLE FROM its parent discipline, nuclear physics. Only a handful of the particles that were to occupy the attention of physicists in the latter half of the century were known. High energies are needed to probe the structure of nuclei and of subnuclear particles and to study their interactions. This is again a consequence of the uncertainty principle: to explore fine detail requires a probe of small wavelength, which implies large momentum, and therefore high energy. For the highest energies, the naturally occurring cosmic radiation was (and continues to be) used as a source. But cosmic radiation is increasingly sparse the higher up the energy scale one goes, so artificial accelerators were already being designed and built in the 1930s. At the very highest energies ever observed (the highest energy yet recorded for a cosmic ray was a staggering 3.2×10^{20} electron-volts—the kinetic energy of a tennis ball at nearly 100 miles an hour concentrated on a single subatomic particle!), cosmic rays are so rare that detectors ranged over 6,000 square kilometers would only be expected to detect a few thousand particles a year with energies above 10^{19} electron-volts. But such an array of detectors is now under construction.[1]

Modest by comparison with the engineering feats of today, the cyclotrons built by Ernest Lawrence were in the vanguard of the new breed of "atom smashers." The largest of these was commandeered during the Second World War, to be used to separate the fissile fraction (^{235}U) of uranium for the construction of the bomb that devastated Hiroshima. Physicists, along with their machines, were redeployed on new missions. When peace came, they had achieved stupendous success, and not only with the new scientific discoveries that had fed the war machine. J. Robert Oppenheimer, who led the Manhattan Project which developed the atom bomb, was profoundly troubled by the work he was doing; other scientists had deliberately set themselves aside from the project. (I know of only one member of the Manhattan Project team who actually resigned from it, and that was Joseph Rotblat, who in 1995 was awarded the Nobel Peace Prize.) The physicists had also learned the skills and methods needed for gigantic collaborative enterprises, designing and engineering new technical devices that both incorporated the fruits of recent discoveries and addressed problems at the frontiers of knowledge. They had acquired courage from the success of their research and audacity in its application. Their science had moved from the laboratory bench to the factories of Oak Ridge and Los Alamos, from a team of a few individuals to the coordinated efforts of hundreds. Big science had been born. One example of this legacy of big science is the ATLAS collaboration, which will use the Large Hadron Collider (or LHC) at CERN, the European laboratory for particle physics (figure 8.1); 1,800 scientists, from 150 institutions in 33 countries, are already participating. Their detector is scheduled to start taking data in 2005 (figure 8.2, plate 15).

The juggernaut of big science, as some would see it, though inaugurated in the exigencies of war, has persisted and burgeoned in peacetime. Huge laboratories have grown up, and not only for high energy physics. CERN, for example, which spans the frontier between France and Switzerland at Geneva, is host at any given time to some 2,000 physicists, not only from its 20 member states but also from dozens of other countries. One of the great rewards for working in this field is to know oneself to be a partner in a truly

Fig. 8.1. Aerial view of CERN. The superposed circles show where the complex of accelerators and storage rings run, underground. (CERN photo. © CERN Geneva)

international enterprise; no flags or other national insignia are permitted inside CERN. It is probably inevitable that all future laboratories of this scale will be international, not only in their scientific personnel but also in their funding. The United States wrote off an expenditure of some two billion dollars and six years of planning and construction when it abandoned its project to build the Superconducting Super Collider in Waxahachie, Texas, leaving behind a useless, incomplete 14.5 mile section of the tunnel which would have housed the machine, and putting hundreds of highly specialized scientists and engineers out of work. The U.S. has now joined the LHC program at CERN.

The discoveries of particle physics in the past half-century have led us to a model of the structure of matter, a set of fundamental particles and the forces which bind them, which has the scope to embrace all the phenomena of particle physics and hence all those of nuclear and atomic physics, and so onward to chemistry and beyond. (Of course this does not mean that there is

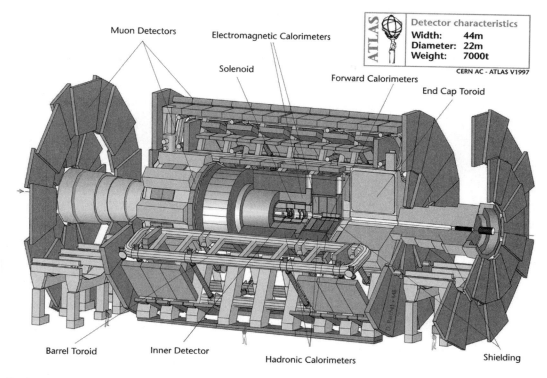

Muon Detectors   Electromagnetic Calorimeters

Solenoid

ATLAS

Detector characteristics
Width: **44m**
Diameter: **22m**
Weight: **7000t**

CERN AC - ATLAS V1997

Forward Calorimeters

End Cap Toroid

Barrel Toroid   Inner Detector

Hadronic Calorimeters

Shielding

Fig. 8.2. A schematic cutaway view of the ATLAS detector. (CERN photos. © CERN Geneva)

nothing to be added as one moves up from one level to the next in the passage from particle physics to chemistry—a climb illustrated in figure 1.2. However, the laws of chemical reactions are, at least in principle, consequences of the properties of the elementary particles, not the other way around.) We have, then, a *standard model* for the fundamentals of particle physics. Standard, because it is accepted by most particle physicists as correct in its essentials. But only a model, because although it provides a marvelous, tightly interlocking framework for the explanation and description of all the observed phenomena, significant loose ends remain. Some predictions, calculable in principle, have so far eluded our technical capabilities. And too many arbitrary parameters still have to be "fed in by hand" for theorists to be comfortable with the standard model as a truly fundamental theory. (One might reflect on the audacious optimism that discomfort reveals. A century ago, most physicists would have accepted that many quantities that entered into their description of the world must simply be accepted as *given*. They could be measured, with ingenuity and accuracy; but it was unreasonable to suppose that they could all be calculated from just a handful of fundamental constants. Now, as a new century dawns, most physicists would agree that it ought to be possible, in principle at any rate, to determine the whole basis of physics from just such a handful of parameters. And the twenty-six or so constants needed to define the standard model are generally regarded as too many!)

Remember that when I speak of particles, I am also speaking of fields: particles are the basic quantum excitations of the fields which bear their names. We may distinguish two different varieties of field. One, of which the electron field is an example, has as its quanta particles which obey Pauli's exclusion principle; no two of them may occupy the same quantum-mechanical state. Electrons, and other particles like them, spin like tiny gyroscopes or tops, with an amount of angular momentum equal to one-half in the natural units of quantum mechanics.[2] Such particles are called fermions after Enrico Fermi, who, along with Dirac, first described their statistics; collections of fermions satisfy a different kind of statistical mechanics from the distinguishable particles of classical mechanics. Another kind of particle, and the photon is one such, carries spin angular momentum of one unit (the spin of the photon is associated with the polarization of light.) Still others have zero spin. Particles with zero or integer spin are called *bosons*, after Satyendra Bose, who together with Einstein described their kind of statistical mechanics.

The behavior of collections of fermions lies at the heart of the structure of atoms and molecules, of the theory of metals and of electronic devices, of the stability of stars. The behavior of collections of bosons underpins the theory of lasers—and the success of Planck's explanation of the black-body spectrum. But physics cannot really be compartmentalized into self-contained subdisciplines, since the exploration of one topic can illuminate the understanding of many others.

For the moment though, let me return to my sketch of the standard model.

It contains fermions, which may be said to describe *matter*; and it contains bosons, which may be considered as the mediators of the *forces*. (See the table at the end of this chapter.) In quantum electrodynamics, for example, the electric and magnetic forces between moving charges are explained through their exchange of *photons*. The matter fields may be grouped together in three sets called *generations*, which are in most respects "carbon copies" one of another, differing only in their masses. (High-energy experiments have determined that there are no more than three generations—a supposition also required to make astrophysical observations consistent with that other standard model of present-day physics, the big bang.) In each generation there is a particle like the electron, carrying one (negative) unit of electric charge; in the three generations, successively, the electron itself, the muon, and the tau. Since these are all electrically charged, they all interact with the electromagnetic field. Associated with each of these particles is a different kind of *neutrino*, one for each generation. The muon is some 207 times more massive than the electron, the tau nearly 17 times more massive yet. These mass ratios are among the parameters of the standard model which have to be "put in by hand." Until 1998 there was no direct indication that neutrinos have any mass at all. (Do not be alarmed by the notion of a particle with zero mass! The so-called *mass* of a particle is really shorthand for its *rest-mass*, related (by $E = mc^2$) to the energy it carries even when at rest. Particles with zero rest-mass can never be at rest; they always travel at the speed of light. Not surprisingly, a photon is just such a particle.) However, physicists had many indirect reasons to believe, indeed to hope, that neutrinos *did* have a mass, and there is now a growing body of evidence to show that they do. Neutrinos are electrically neutral and have no interactions with the electromagnetic field. But like *all* the matter particles, they do have what are called *weak interactions*; indeed, neutrinos have no other kind, except gravity. The weak interactions are so feeble that neutrinos generated in the thermonuclear reactions that power the sun pass freely through the earth, with barely any chance of interacting at all. Yet the rare interactions of solar neutrinos have been detected and studied, and also at higher energies, neutrino beams have been generated and deployed. Neutrinos were first directly detected in a nuclear reactor in 1956, by Frederick Reines and Clyde

Cowan. Reines had described the search for the neutrino as "listening for a gnat's whisper in a hurricane," but as the *Los Angeles Times* said in his obituary: "He heard, and altered the view of the universe."

The Super-Kamiokande experiment, which has shown that at least one of the neutrinos has a mass, illustrates the subtleties of particle physics. Although neutrinos interact exceedingly feebly, they are produced quite copiously by cosmic rays entering the atmosphere: about 100 such cosmic ray–induced neutrinos pass through you every second. Yet the chance that one of them will have undergone an interaction as it does so is only around one in ten in your entire lifetime! To study such rare interactions requires a massive detector, in this case a tank containing 50,000 tonnes of ultra pure water in a zinc mine 600 meters below Mount Ikena in Japan (figure 8.2). The overlying rock acts as a barrier to most of the cosmic radiation but is essentially transparent to the neutrinos. Indeed, the crux of the experiment is to compare the rate of detection of neutrinos produced in the atmosphere overhead with those coming up through the detector, having been produced on the other side of the earth and then passed through it. Remember that 100 neutrinos a second pass through your body, so around 30 million a second pass through the huge detector (figure 8.3). Occasionally, five or six times a day, one of them collides with an oxygen nucleus in the water and produces either an electron or a muon. These charged particles, with enough energy to be traveling at close to c, the speed of light in a vacuum, are in fact traveling faster than the speed of light in water, which is about 3/4 c. There is no conflict with the theory of relativity, which forbids a particle from moving faster than c, the speed of light in a vacuum. Just as a projectile which exceeds the speed of sound produces a shock wave, the sonic boom, a charged particle traveling faster than the speed of light in a transparent medium produces a sort of shock: light emitted in a cone which then illuminates the wall of the tank in a faint ring of so-called Cherenkov radiation. From the details of that flash, lasting only a few billionths of a second, which is detected and measured by the huge array of photomultiplier tubes (there are 11,146 of them, each 20 inches in diameter, each needing its own electronic readout) which line the walls of the tank, the properties of the particle produced can be determined. The ratio of electrons to muons produced in

Fig. 8.3. The Super-Kamiokande detector, here seen half-filled with water while under construction. (Institute for Cosmic Ray Research, The University of Tokyo)

these collisions varies with the direction of the incident neutrino, and this variation gives us the clue to neutrino mass. For those coming from overhead, the ratio is higher than for those coming from below. All these neutrinos have originated in cosmic ray collisions in the atmosphere, so those coming from below will have been produced 13,000 kilometers away, on the other side of the earth. The only explanation consistent with the data is that some of these neutrinos, in the course of their passage through the earth will have turned into something else.

An exhaustive analysis of the data from Super-Kamiokande and other experiments[3] leads to the conclusion that the muon neutrino *oscillates* with something else, in all likelihood a tau neutrino, so that there is a regular transition from muon neutrino to tau neutrino and back again. For those coming from overhead, too little time elapses for this to have made an appreciable diminution in the flux, but in the case of those coming from below, sufficient time passes so that what started off as a muon neutrino will have only a 50 percent chance of being one when it reaches the detector. This kind of oscillation is known from elsewhere in particle physics; a particle called the neutral kaon also displays this specifically quantum kind of behavior. More precisely, there are two different kinds of neutral kaons which oscillate back and forth from one kind to the other. And such oscillations can only occur when there is a difference in mass between the states involved, which determines the rate at which the oscillations take place. Now you cannot have a mass *difference* between massless particles. Hence the inference that neutrinos have non-zero mass. All that can be determined so far is a bound on the mass difference, and so on the mass of the lightest of them. This mass difference is found to be greater than 0.07 in the units (eV/c^2) used by the physicists (in these same units, the mass of the electron, hitherto considered the lightest of the fundamental particles, is 510,999). You should not think that neutrinos are some rare and exotic kind of matter. For every electron in the universe, there are a billion neutrinos, and in every cubic meter of space, 300 million "relic" neutrinos left over from the big bang. So the fact that neutrinos have a mass means that they can contribute significantly to the total mass of matter in the universe. A mass of 0.07 eV/c^2 would imply that the total mass of the neutrinos in the universe is comparable to that of the total mass of the visible stars! This has important consequences for cosmology.[4]

Collectively, the electron and its look-alikes in the other two generations, and their cousins the neutrinos are called *leptons*,[5] a name first introduced by Abraham Pais and Christian Møller. The remaining matter particles in each generation are the *quarks*. These are the building blocks for the protons and neutrons of which atomic nuclei are composed, and all the other members of that broad family of particles called *hadrons*,[6] the name for particles that have *strong interactions*. Protons, neutrons, and the like, are no longer regarded as primary, fundamental particles. Rather, they are composites, a proton being built from three quarks.[7] The quarks carry two different properties, called *flavor* and *color*. (Of course, these are just names, and not in any way to be taken literally!) In each generation a pair of flavors is present; *up* and *down* in the first generation, *charm* and *strangeness* in the second, and *top* and *bottom* in the third. These rather curious names arose before the model was fully elaborated. In the case of the up/down pairing, it refers to isospin. Isospin, mathematically analogous to ordinary spin, is a kind of abstract symmetry originally introduced into nuclear physics by Heisenberg. Just as the ordinary spin of a particle with spin-½ can be described in quantum mechanics in terms of just two distinct basic states, which we might just as well describe as the spin axis pointing *up* and spin axis pointing *down*, so also there are two distinct basic states for isospin-½, again called *up* and *down*. But the isospin symmetry has no direct connection with ordinary space; it refers to an abstract sort of internal space useful in describing the particle. The theory of high-energy physics abounds with such "internal symmetries."

Symmetry plays a profound role in high-energy physics, both aesthetic and practical. There are countless manifestations of symmetry in nature, for example, the (approximate!) left/right symmetry of our own bodies or the pristine symmetry of crystals. Some, like these, are manifested in a way we can see directly, as with mirror images or the discrete rotation and translation symmetries of the pattern on a wallpaper or a tiled floor. There are other, more abstract kinds of symmetries, best described in mathematical terms, which help to classify and to explain the physics of atoms and the subnuclear world of particle physics.[8] These symmetries often allow us to recognize that two or more particles which might at first seem quite distinct are in fact different manifestations of the same thing. It's rather like recognizing that the symbols ≻ ≺ ∧ ∨ are related one to another by symmetry, and so can be regarded as different manifestations of the same shape.

There are symmetries acting *between* generations, rather than within them, but these are not regarded as fundamental. One such symmetry is the celebrated SU(3) of the "eightfold way" proposed by Gell-Mann, which substantially simplified and gave order to the phenomenology of hadrons. Such symmetries are consequences of more basic properties of the standard model. The quarks have masses, which are again inputs to the model, put in "by hand." The "top quark," in fact, is so very massive that its production is only just within the scope of the most powerful accelerators so far constructed; its existence was only confirmed in 1995, and it has a mass greater than that of a whole atom of tungsten (plate 16). Perhaps the greatest challenge confronting particle physics at the turn of the century is to understand why the masses of the quarks vary so much: the "top" quark is some 35,000 times as massive as the lightest, the "up" quark. Discovering the origin of these masses is one of the principal goals for the first experiments proposed of the LHC.

The quarks carry electric charge; in units in which the charge of the electron is -1, the up, charm, and top quarks have charge $+\frac{2}{3}$, while the down, strange, and bottom quarks have charge $-\frac{1}{3}$. They therefore all experience electromagnetic interactions, and like leptons, they also participate in the weak interactions. These two kinds of interaction have very different properties: electromagnetic interactions are many orders of magnitude larger than their weak counterparts and forces fall off slowly with separation, while the weak interactions, in addition to being feeble, have a very short range. But remarkably the two are but different aspects of the *same* unified interaction, the *electroweak interaction*. The unification of electromagnetism and the weak force earned Nobel Prizes for the Pakistani physicist Abdus Salam and the Americans Steven Weinberg and Sheldon Glashow. It means that the photon, as the mediator of electromagnetism, and the spin-1 bosons associated with the weak interactions are siblings. That means that at very short distances particle behavior departs from the predictions of quantum electrodynamics. There are subtle consequences which arise even for atomic physics, and such effects have indeed been measured.

That this unification of electromagnetism and the weak interactions is possible, despite the masslessness of the photons and the massiveness of the weak-interaction bosons (they are nearly 100 times as massive as a proton!), is due to a miracle known as *spontaneous symmetry breaking*. At high energies, a symmetry between electromagnetism and weak interactions is believed to be manifest explicitly: we would observe no distinction between them, and the bosons mediating them would all, like the photon, have zero mass. But at lower energy, the symmetry is spontaneously broken in a phase transition analogous to that which occurs when a liquid solidifies as it is cooled. For example, water is isotropic; that is, no direction is singled out in the arrangement of its molecules. But in crystals of ice, they are arranged in regular planes which, because they are aligned all in the same direction, have broken the rotation symmetry present at higher temperatures. The consequence of a similar phase transition in the electroweak field is to break, not a symmetry in space, but an internal symmetry in the interactions, and in the process the bosons associated with the weak interactions acquire a mass, and the effective strength of the weak interaction becomes small. Detailed predictions follow from this scenario, which are spectacularly confirmed by experiment, not least the existence of the massive bosons mediating the weak interactions, their exact masses, and other properties. These bosons are called the W and the Z and were discovered in 1983 in experiments at CERN led by the Italian Carlo Rubbia. These experiments relied on achieving effective energies for particle collisions far higher than those previously attained by bombarding a fixed target with a beam of protons. Instead high energy beams of protons and antiprotons were made to collide head-on, a feat made possible by a technique for concentrating the beams developed by the Dutch physicist-engineer Simon van der Meer. Rubbia and van der Meer shared the 1984 Nobel Prize for their achievement. Rubbia led a team of around a hundred physicists, driving them with ferocious energy while making regular transatlantic flights to meet his teaching commitments at Harvard.

The quarks, as befits the building blocks of hadrons, have *strong interactions* too. The spin-1 bosons which mediate these strong interactions are called *gluons* (the younger generation of physicists were more banal than their elders in dreaming up names for particles!)—they glue together quarks inside protons, neutrons, and the other observed hadrons. The subtleties of this part of the model are really marvelous. A proton, for example, is composed from two up quarks and a down quark. And yet for this construction to be possible, along with the observed properties of the proton and the other hadrons associated with it, implies something most unusual about the quarks. For each of the six different flavors of quark we have so far introduced (up, down, etc.), there are three different *colors*, so that there are three up quarks, differing only in their color, three down quarks, and so on. It then turns out that the whole structure fits together, and although the quarks, as previously mentioned, all have fractional charges, the resultant hadrons which are made from them only have integer charges. The key to all this is the participation of the gluons; they interact only with color, and so the leptons, being colorless, do not experience the strong interactions at all.

Another feature of the subnuclear world is also of crucial importance for the model. There are eight different kinds of gluon, which themselves have color and so interact with one another. Because of the particular way that they interact, the force between colored particles, quarks, or gluons *increases* with increasing separation, and conversely becomes weaker at short distances. This is like no force encountered before. As a consequence, color is *confined*, so that we never observe a free quark detached from others, nor a free gluon. None of the particles directly observed carry color, because the colors of their constituents cancel out. In consequence, none of the particles we observe directly carry fractional electric charge, as do the quarks. Although they are not detected, and perhaps are undetectable, as free particles, the experimental evidence for the existence of quarks and gluons with the properties I have tried to summarize is overwhelming. The theory of strong interactions based on quarks and gluons has many formal similarities to QED, the tried and trusted starting point for relativistic quantum field theories. It is called QCD, quantum chromodynamics.

My description of the standard model is now almost complete. To recapitulate, *matter* is organized into three generations, which apart from very different masses are almost carbon copies of one another. In each generation there is a pair of leptons and three pairs of quarks, and all eight of these particles are spin-$\frac{1}{2}$ fermions. I should add that for each of them there is a corresponding antiparticle, with opposite flavor and opposite color. The *forces* are associated with spin-1 bosons: the strong force with the eight massless gluons, electromagnetism with the photon, also massless, and the weak force with (let me now name them) the W^+, W^- and Z^0 particles.

I fear that my very abbreviated outline of the model made it all sound very ad hoc. To be sure, some of its features were introduced so as to fit the facts, to "save the phenomena" (for example, the number of generations or the masses of the quarks). What I have not been able to convey, for that would have required considerable mathematical elaboration, is that in fact the whole scheme is very tightly organized and admits very little scope for adjustment. The key to this is the nature of the forces, the bosons which carry them, and the fields of which they are the quanta. They are *all* associated with *gauge symmetries*, and as was already mentioned, it is just this property which ensures that the theory is *renormalizable*. The underlying symmetry is in fact very simple,[9] so much so that there are only three constants needed to fix the strength of *all* the interactions.

It also happens that with three generations it is possible to incorporate in a simple way the observed breaking of parity (left/right) symmetry and the tiny effect which is responsible for the asymmetry in the universe between matter and antimatter. So far as the first effect is concerned, in the absence of this symmetry breaking, the fundamental interactions would proceed in exactly the same way for any given system as for its mirror image. And indeed, for most everyday situations this is so; if I can construct a machine to make right-handed screws, I can build its mirror image, which will make left-handed screws just as well. But there are left/right asymmetries in nature. The most familiar one is the preponderance of right-handed individuals in the human species, but this is probably just the outcome of an accident in the course of evolution. Perhaps more significant is the fact that many of the molecules involved in the metabolism of living creatures are "handed"; that is, the mirror image does not have the same properties as the original. All the amino acids in living organisms are left-handed. The sugar dextrose differs from its mirror image counterpart levulose; we metabolize dextrose with ease, but not its right-handed twin. Cosmic beta radiation is polarized, and some evidence suggests that it could preferentially have destroyed right-handed amino acids. So it is just possible that the preponderance of one-handedness in the molecules in our body results from the tiny asymmetry in the dynamics associated with parity violation. It is amusing to note that the molecules which make oranges taste different from lemons are the mirror images of one another. In a world without parity violation, would oranges and lemons taste alike?

The (broken) parity symmetry is often denoted *P*, and there are two other symmetries which, like P, are broken by the weak interactions. One is C, charge-conjugation; this is respected by the strong and electromagnetic interactions which are unaltered if every charge is replaced by its opposite. Finally there is T, which is rather more subtle. T stands for time reversal, and time-reversal symmetry means that for every possible process, another is possible, described by the same quantities but with the arrow of time reversed. In classical physics this could be thought of as running a film of the process backward; if the first film shows a billiard ball collision, so also will its time-reversed version—and the same laws of physics apply to both. One very profound consequence of quantum field theory is that the combination CPT is always an unbroken symmetry. That both C and P symmetries were separately broken by the weak interactions was postulated by the Chinese-American theorists Chen-Ning Yang and Tsung-Dao Lee in 1956 (they were awarded the Nobel Prize for this work in the following year) and was first verified by Chien-Shiung Wu, another Chinese American, and her collaborators shortly thereafter.

Nevertheless, the combination CP, and therefore also T, might still be left unviolated. But in an experiment led by James Cronin and Val Fitch at Princeton in 1963—the results were only published the following year after prolonged analysis to check its startling conclusion—it was shown that even these symmetries are broken, albeit very weakly. (This also led to a Nobel Prize, shared by Cronin and Fitch in 1980.) Their experiment involved subtle quantum mixing effects manifested by neutral kaons.[10] Measuring the parameters of CP violation is difficult, and new experiments are planned that will pin them down more precisely at detectors designed with this as their principal objective. One is called BaBar, for B$\bar{\text{B}}$, and it began taking data in 1999 (figure 8.4). It will be used to study quantum mixing between B-mesons and their antiparticles $\bar{\text{B}}$, which are analogous to the K and $\bar{\text{K}}$, the neutral kaons studied by Cronin and Fitch.

I have left until last one further essential, missing ingredient which is needed to make the whole of the standard model cohere. Mention has already been made of *spontaneous symmetry breaking*. The symmetry which governs the electroweak interactions would, at face value, seem to imply that all four of the bosons involved, the photon and the W^+, W^-, and Z^0, should be

Fig. 8.4. BaBar, the detector at Stanford University's Linear Accelerator Laboratory. (Photograph by Joe Faust/ SLAC)

massless and interact in the same way, with the consequence that the weak interactions would not be weak and would not have their observed very short range. The missing ingredient is another particle, a spin-0 boson. This enters the theory in such a way that although the fundamental structure has the high degree of symmetry alluded to, this fully symmetrical structure is inherently *unstable*. It is predicted that at very high energies the symmetry does become manifest. Indeed, in the very early universe, when the enormously high temperatures associated with the big bang of creation caused particles to collide with one another at very high energies, it *was* manifest. But at the relatively low energies of even our most powerful accelerators, the symmetry is broken and the instability drives the interactions to a less symmetrical but more stable configuration. This possibility, and the role of the spin-0 boson associated with it, was first described by Peter Higgs. The search for the Higgs boson, the missing piece in the jigsaw puzzle, is one of the prime motives for pressing to the next generation of particle accelerators, which will start collecting data early in this century. Just before it was shut down in November 2000 to allow construction of the LHC to commence, there were tantalizing glimpses of what *might* be the Higgs boson reported from CERN's LEP (Large Electron Positron) accelerator. We do not know the mass of the putative, elusive Higgs boson; but experimental constraints and theoretical prejudice, and perhaps an element of wishful thinking, lead one to anticipate that its production should be possible at the LHC, the Large Hadron Collider under construction at CERN, and it should then be seen by such giant detectors as the 7,000-tonne mammoth of the ATLAS collaboration—as big as a five-storey building. Not only will the detection of the spin-0 boson confirm the standard model, with its harmonious and orderly predictive capabilities; it will also shed light on the origin of *mass*. For the interactions of the Higgs boson are intimately related to the masses of the particles, and it is the study of these interactions that may be expected to dominate the high-energy experimental program of the first decades of the twenty-first century. One way or another they will surely yield clues to solve the mystery of mass for it is through their interactions with the Higgs boson that the particles of the standard model acquire mass when the symmetry is spontaneously broken.

In order to create the Higgs boson, particle collisions at very high energies are required, and in such collisions very many other, more familiar particles will be created too. From a typical collision it is calculated that hundreds of particles will be created, and something like a billion such collisions will take place each second! To track down these particles, to determine their kinematics and explore just what happened in each particular collision—hoping to find one in which the elusive Higgs particle was created—is a phenomenal challenge. The flow of data from the detectors will be as great as that from three billion music CDs played simultaneously, and from that cacophony just the one significant melody of the Higgs boson must be selected. The fast electronics needed to achieve this have only very recently been developed, but will doubtless have applications beyond the world of high-energy physics.

After all, previous advances in electronics and data processing for high-energy physics have found their way into fields as diverse as medicine and manufacturing, entertainment and telecommunications.

For all its organizing capacity, for all its tight economy, few assumptions and many predictions, the standard model is incomplete. For one thing, it points to a wider unity, between the strong interactions and the already unified electroweak interactions. Extrapolating from the way the strength of interactions varies as the energy of collisions is increased, we have reason to believe that such a unification might indeed occur. There are grand unified theories (GUTs) waiting in the wings, ready to be applied should the evidence become more compelling. GUTs play a significant role in present-day cosmology, to which we will turn in chapter 11.

Another possible unification, still more daring, but nevertheless hinted at by experiment and attractive to theory, is one between matter (as represented by the fundamental fermions) and force (the bosons). Such a *super-symmetry* would require a whole range of new particles, sleptons and squarks, photinos and gluinos. None of these has yet been detected, presumably because all of them are too massive to have been made at present-day accelerators. However, an extrapolation from what is measured at the highest energies so far explored suggests that grand unification *does* occur, and at the same time lends credibility to the supersymmetry hypothesis. Is this all just some exotic fantasy? Perhaps; but in a later chapter we shall see that supersymmetry is needed as an ingredient to make strings into superstrings, and thereby to generate what is widely believed to be the *only* kind of theory which will at last admit a consistent quantum theory of gravity. Notwithstanding its far-reaching success in describing the fundamental particles of matter and the forces between them, the standard model is incomplete. No fundamental account of matter and its interactions can exclude gravity, and the standard model makes no mention of this universal force. To go beyond the standard model is one of the challenges addressed by superstring theory.

THE PARTICLES OF THE STANDARD MODEL

The Ferminons

Generation	First	Second	Third
Quarks	u	c	t
	d	s	b
Leptons	e	μ	τ
	ν_e	ν_μ	ν_τ

The quarks are designated by their initials: up, down, charm, strange, top, bottom. Each kind comes in three "colors." The leptons are respectively the electron, the muon, and the tau, and their corresponding neutrinos.

The Bosons

Interaction	Force-carrier
Electroweak	$W^+ \ Z^0, \gamma \ W^-$
Strong	g

There are eight different gluons (denoted *g*), which interact with the quarks and with one another. The W^+, Z^0, W^-, and γ (photon) interact with the different kinds of electroweak charge.

Finally, standing on its own is the Higgs boson (H).

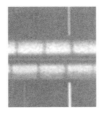

9

WEIGHTY MATTERS

★ ★ ★

The General Theory of Relativity

WHAT IS GRAVITY? NEWTON RECOGNIZED "THAT THERE IS A power of gravity pertaining to all bodies, proportional to the several quantities of matter which they contain," which varied "inversely as the square of the distance." But as to its cause, its real nature, he was deliberately noncommittal: "But hitherto I have not been able to discover the cause of those properties from phenomena, and I frame no hypotheses; for whatever is not deduced from the phenomena is to be called an hypothesis. . . . And to us it is enough that gravity does really exist, and acts according to the laws which we have explained, and abundantly serves to account for all the motion of the celestial bodies, and of our sea." (He refers here to his explanation for the tides.)

And so it remained: a great deal of metaphysics and much first-rate application in astronomy and in other areas of physics, but very little of consequence by way of fundamental theory, until 1915. It was in that year that

Einstein published his most radical achievement, the general theory of relativity.

The force of gravity is what we experience as *weight*. Never mind that the story of Galileo dropping different masses from the Leaning Tower of Pisa is probably apocryphal; one way or another, he did indeed discover that the force of gravity produces identical accelerations on different masses. As a consequence, when a spacecraft is in free flight, falling freely round the earth, all its contents, being likewise in free fall, accelerate in the same way. (Yes, the orbit of the spacecraft is in fact *free fall*; the gravitationally induced acceleration toward the center of the earth is precisely that needed to keep the craft in its circular orbit, just as the tension in a string with a stone tied to its end keeps the stone whirling round in a circle.) Weight as we perceive it is the force attracting us to the center of the earth, pulling us downward. Without opposition, that force would indeed accelerate us downward, as when we jump or fall. To prevent this, to hold us in position, we need an opposing force, such as the pressure on the soles of our feet or the compression of the spring in a weighing machine. It is this opposing force which we commonly use to measure weight; so when, as in the spacecraft, no such force is needed to keep its occupants at rest relative to their surroundings, we say they are *weightless*. It is not that the force of gravity, their weight, has been eliminated; it is simply that the usual consequences of that force (the pressure on the soles of their feet or the fall to the floor) have been canceled out by the acceleration to which they as well as the spacecraft itself are subjected as they fall together around the earth. In just the same way, your weight seems to change in an elevator when it accelerates, speeding up or slowing down.[1] Stand on scales while riding on an elevator, if you need more evidence than the feeling in the pit of your stomach!

In trying to grapple with the problem of gravity, Einstein, characteristically, placed these simple observations at the center of his theory. Already in 1907, while still employed at the Patent Office in Bern, he was struck with the thought, "If a person falls freely, he will not feel his own weight." This is the germ of the *principle of equivalence*, which asserts that gravitational forces are indistinguishable in their effects from those experienced when the reference frame (elevator, spacecraft, lurching train) is accelerated. His goal was now set: to remove the restriction he had imposed in his special theory, namely, that it related only to the description of events by observers in relatively *unaccelerated* motion, in which case any two observers would agree that, in the absence of imposed forces, bodies continue to move with constant velocity. He now wanted to extend his transformation laws so as to be able to deal with more general relative motion between different observers. The principle of equivalence already suggests that the resultant theory will have to deal also with gravity, and in fact will put gravitational forces in a privileged category.

Einstein also recognized that any theory general enough to encompass relatively accelerated observers would have consequences for the understanding of space and of time, and gave a simple illustration of why this must be so. A particular case of accelerated motion is motion in a circle. Any motion other than with constant *velocity* (that is to say constant speed in a straight line) is accelerated. Moving in a circle, even at constant speed, implies acceleration, since the direction of motion is changing. Suppose that an observer at the center of a rotating disc makes measurements to determine its diameter; the result will be the same as though the disc were stationary. But if now the observer on the rotating disc measures its circumference, it will be found to be less than for a stationary disc. This is a consequence of the Lorentz-Fitzgerald contraction, which shrinks lengths along the direction of motion. But it leaves the measurement of the radius unaltered, since the velocity of every portion of a radius is perpendicular to its length. The result is that the perceived ratio of the circumference to the diameter will be *less* than if the disc were at rest. So the observer of a rotating disc will find that the ratio of circumference to diameter is not π, which is what Euclidean geometry requires! The same observer will also find that clocks placed at the circumference of the disc run slow compared with one at the center. There is no way to set up on a rotating disc an overall spacetime description of events which conforms to the precepts of special relativity, still less to those of Newtonian physics and Galilean relativity: rulers appear to contract and clocks to slow down as they are taken from the center to the circumference of the disc. In fact, it is not possible for any observer, from any point of view, to give a description that holds good for the whole disc and is consistent with Euclidean geometry and the data from synchronized clocks. An observer on the rotating disc, however, experiences a centrifugal force, and the principle of equivalence tells us that this should have the same effect as an identical gravitational force. "We therefore arrive at the result: the gravitational field influences and even determines the metrical laws of the space-time continuum. If the laws of configuration of ideal rigid bodies are to be expressed geometrically, then in the presence of a gravitational field the geometry is not Euclidean."[2]

Clearly, Newtonian gravity must somehow find its place in the new theory, and so it does, albeit as an approximation valid when the gravitational accelerations are not too great nor relative velocities too high. The source of Newtonian gravity is *mass*, the mass of the idealized pointlike particles that Newton considered to be the ultimate constituents of matter. When generalized to take account of distributed mass, rather than pointlike concentrations of mass, the source of gravity in Newtonian theory becomes mass density, the ratio of mass to volume. Since Einstein's special theory of relativity had already shown the interconvertibility of mass and energy ($E = mc^2$ again!), in a relativistic theory of gravity, energy density might be expected to figure as a source of gravity. But in special relativity, energy density does not enter the theory on its own; it is part of a complex of ten quantities (including, for example, momentum density) which transform together. Their values will differ for two observers in motion relative to each other, but the transformation rules from one observer to another are quite simple and are characteristic of what is called a *tensor*. So Einstein wanted to construct a theory of gravity which had a tensor as its source. He also wanted a theory in which the laws of physics could be expressed in the same way no matter what the relative motion of the observers in whose frame of reference those laws apply. (In mathematical terms, this imposes the technical requirement that the laws of physics should be *covariant* under the most *general* coordinate transformations.)

Einstein was now shifting his attention to a new question: What does it mean to change the frame of reference, to change the way we label the events of spacetime? The special theory had accommodated only a restricted class of such changes, but now Einstein grappled with coordinate changes in

Fig. 9.1. Greenland appears to be bigger than Africa in this Mercator projection map. (Courtesy of Graphic Maps, Galveston, Texas)

general. René Descartes had shown how convenient it was to imagine a plane as ruled with a grid of perpendicular lines, and the cartesian coordinates so introduced are familiar in graph paper and maps. A point on a map, and its counterpart on the surface of the earth, are identified by a pair of numbers, latitude and longitude. On a large-scale map covering only a small area, the north-south lines and the east-west lines are indeed well approximated by orthogonal sets of straight lines; but as Mercator and the other great mapmakers well understood, it is not possible to map the whole surface of Earth, nor even a large fraction of it, without distortion. Greenland may appear to be as large as Africa (figure 9.1)! This means that the actual distance between two points on Earth is not just proportional to their separation on the map, but corrections are needed, corrections which depend on location. The shortest route between two points on Earth will not map into a straight line; great circle routes on an airline map appear to be curved. These properties of geographical maps encode the fact that the surface of Earth is *curved*; it has the intrinsic curvature of the surface of a sphere. Unlike the surface of a cylinder (which can be cut along a line parallel to its axis and flattened out into a plane without any stretching or distortion), the surface of a sphere can only be mapped onto a plane with some degree of distortion. Einstein wanted to consider the general transformations between different labeling—different coordinate choices—of the points of spacetime, transformations that were of this kind. In his theory spacetime possesses intrinsic curvature. This is encoded by the way that the separation between events is determined from the difference in their coordinates (space and time labels), just as the intrinsic curvature of Earth is encoded in the mapmakers' rules for determining distances from the separations of longitude and latitude.

What then determines the curvature of spacetime? The mathematical description of curvature had been initiated by Karl Friedrich Gauss,[3] and extended from his study of two-dimensional surfaces to consideration of higher dimensions by Bernhard Riemann in the famous "trial" lecture (*Probervorlesung*) he delivered at Göttingen in 1854, before his appointment to a much-desired, though unpaid, lectureship at that university. The tensor calculus developed for this purpose by Riemann and others was by no means part of the mathematical baggage carried by theoretical physicists in the first decades of the century. Although Einstein recognized that it provided the key to his problem, he encountered many pitfalls and went down many false trails in his struggle to find the way to ensure general covariance and was all but defeated by the mathematical difficulties involved. It took eight years before the final result was published as "The Field Equations of Gravitation" in 1915. In 1919, at the end of the introduction to a paper in which he had presented a more restricted version of the theory, he wrote: "The magic of this theory will hardly fail to impose itself on anybody who has truly understood it; it represents a genuine triumph of the method of absolute differential calculus, founded by Gauss, Riemann, Christoffel, Ricci and Levi-Cività." Within a year he had published a comprehensive formulation of the theory, and had already determined three principal predictions it made that differed from those of Newtonian physics.

The first of these is the *gravitational redshift*, which can be thought of as follows. A photon has energy proportional to the frequency of the light in question. If it goes "uphill," that is, against the pull of gravity, it will lose energy, and the frequency of the light will be reduced, which means that the color of the light will shift toward the red end of the spectrum. For this reason, the characteristic frequencies of lines in the spectrum of sunlight are slightly redder than the corresponding lines from a source on Earth, since to reach us from the solar surface, the light has to climb away from the massive gravitational attraction of the sun. The shift is only about four parts in a million and is hard to distinguish from other effects that also can cause small changes in the observed frequencies, so more clear-cut observations of the gravitational redshift had to wait for new technologies that could measure the very much smaller result predicted for a terrestrial experiment. This was first achieved in the late 1950s, when Robert Pound and Glen Rebka measured the change in frequency (by only two parts in 10^{15}!) between the top and bottom of a 22-meter tower at Harvard University. The shift was exactly as predicted nearly half a century earlier.

In the sixteenth century, Johannes Kepler had deduced from the amazingly accurate and detailed records made by Tycho Brahe on the motions of the planets across the heavens that those "wandering stars" went round the sun on *elliptical* paths. (Tycho was a Danish nobleman; he had no telescopes but did have an observatory, with quadrants, sextants, and other instruments, at his castle, Uraniborg. The observatory was funded by his estates and the tolls from the Baltic trade supplemented by pensions and gifts from the Danish crown; the total cost represented a fraction of the national GDP similar to that of Denmark's CERN subscription today.) Newton's theory of gravity explains the elliptical orbits of the planets, and even the small corrections consequent on the fact that each planet disturbs the motions of all the others. The effect on Mercury is the most pronounced; it is the lightest of the planets, with a rather eccentric orbit which it completes in just 88 days. Its perihelion (the point at which it approaches closest to the sun) is in a slightly different direction from the sun from one circuit of its orbit to the next; this gradual change in direction is called the advance of the perihelion. This was known to Urbain Le Verrier in the 1850s and was determined with greater precision by Simon Newcomb in the 1880s. It amounts to some 5,599 seconds of arc a century. The Newtonian corrections account for 5,556 seconds of this change; general relativity accounts for the remaining 43. This was the second of the "classical tests of general relativity."

The gravitational redshift already shows that light is affected by gravity, so perhaps it is not so surprising that light rays are deflected by gravity. A ball arcs in a parabolic curve as it falls in the earth's gravitational field; the planets fall around the sun in their elliptical orbits; so why should not light rays from distant stars be deflected if they pass close to the sun (where its gravitational pull is strong) on their way to us? A simple calculation can be made, treating light in the same way as matter, which led to such a prediction; Einstein's theory of 1915 confirmed that such an effect is to be expected but gave *twice* the value of (his) earlier calculation. A moment's consideration shows that it is not easy to measure this; for one thing, the effect is small (even for rays which just "graze" the edge of the sun as they travel from a star to us, it is only 1.75 seconds of arc). But in any case, to see the stars in a direction close to that of the sun, the observations have to be made in daytime—when the brightness of the scattered light from the sky makes the stars invisible! However, stars do appear in the daytime during total eclipses of the sun. Einstein's paper appeared in 1915, when Europe was engaged in less peaceful activities than star gazing. Prompted by Einstein's earlier calculation, an expedition of German astronomers had set out in 1914, just before the outbreak of the war, to observe a total eclipse in Russia. Einstein wrote to Paul Ehrenfest, "Europe, in her insanity, has started something unbelievable. . . . My dear astronomer Freundlich [the leader of the expedition] will become a prisoner of war in Russia instead of being able there to observe the eclipse of the sun. I am worried about him." The members of the expedition *were* held for a short time as prisoners of war, before being exchanged for some Russian officers.

Another total eclipse was predicted for 1919. Sir Arthur Eddington, one of the leading astronomers of the day, was a Quaker, moved to try to repair the rifts between British and German science which had opened during the war. Already in 1917 he was laying plans to have observations made both from Brazil and from an island in the Gulf of Guinea, off West Africa. In his book *Space, Time and Gravitation*, he wrote:

In a superstitious age a natural philosopher wishing to perform an important experiment would consult an astrologer to ascertain an auspicious moment for the trial. With better reason, an astronomer of today consulting the stars would announce that the most favourable day of the year for weighing light is May 29. The reason is that the sun in its annual journey round the ecliptic goes through fields of stars of varying richness, but on May 29 it is in the midst of a quite exceptional rich patch of bright stars—part of the Hyades—by far the best star field encountered. Now if this problem had been put forward at some other period of history, it might have been necessary to wait some thousands of years for a total eclipse to happen on that lucky date. But by a strange good fortune an eclipse did happen on May 29, 1919.[4]

The clouds which had obscured the sun dispersed in time for the measurements to be made—and the photograph taken during the few minutes of total eclipse revealed that the stars appearing close to the sun were indeed displaced from their positions relative to those further away, just as general relativity predicts, that is, twice as much as the earlier "Newtonian" calculation. It should be appreciated that observations of this kind are difficult and subject to errors, so that even if one takes into account the observations made during other solar eclipses in the past eighty years, a skeptic might still be unconvinced that the theory was confirmed by these results. However, it is now possible to make analogous measurements using radio telescopy to observe the occultation of a quasar, without having to wait for a total eclipse. And the precision of agreement between theory and observation is better than 2 percent.

To the three "classic" tests there has been added a fourth, proposed and conducted by Irwin Shapiro in the 1960s. The time taken for light to travel is affected by gravity; again, the curvature of spacetime caused by the sun has a measurable effect. Using very high-powered transmitters and sensitive detectors, it is possible to record the echo of a radar pulse reflected from a planet and to track the change in the time delay between transmission and detection of the echo as the planet passes in close conjunction with the sun. Again theory and observation agreed in a satisfactory way, but still not so overwhelmingly as to sway a persistent skeptic. The observational basis for accepting general relativity was good, but the real support for the theory came from its internal logic and self-consistency.

This has changed in the last few years. Through observations spanning

Fig. 9.2. The radiotelescope at Arecibo, built in the crater formed by a giant meteor impact. It is the largest single-dish telescope in the world. (Photograph by David Parker. Courtesy of the NAIC-Arecibo Observatory, a facility of the NSF)

over twenty years of a system discovered in 1974, the theory of general relativity was subjected to tests of much more vigorous accuracy than ever before and can now be asserted to have passed—with a grade of A. These observations are of a *binary pulsar*, discovered by Joseph Taylor and Russell Hulse of Princeton University, working at the Arecibo radio telescope (figure 9.2). Their study of the binary pulsar, and the conclusions they drew from it, won them the 1993 Nobel Prize.

As said before, there are many binary stars, two stars which orbit one another. In some cases, the plane in which the stars orbit is close enough to our line of sight that the stars eclipse one another regularly. More commonly, the binary system is identified from the regular periodic shifts in the spectral lines of one of its members, shifts caused by the Doppler effect as the star

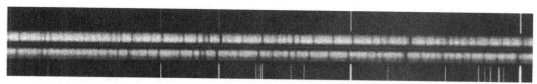

Fig. 9.3. The orbital motion of a binary star, revealed by the Doppler shifts in its spectrum. This is a portion of the spectrum of Castor (one of the "heavenly twins"). The bright vertical lines at top and bottom are reference emission lines produced in the spectrograph; the middle two bright spectra crossed by dark, vertical absorption lines are spectra of the star on two dates. Offsets of these lines, first to the red and then to the blue, show that the star is receding and then approaching, because it is orbiting around another star. (Observations made at the Lick Observatory, Mt. Hamilton, California)

moves now toward us and then away along its orbit (figure 9.3). When a star has exhausted all its thermonuclear fuel, it collapses under its own weight; if it is not too massive, it will become a *white dwarf*, as small as a planet, glowing like a fading ember. A more massive star will collapse still further, to become a *neutron star*, a tiny ball just a few kilometers across into which all the mass of the star is compressed. An isolated neutron star would be all but undetectable, were it not for the fact that in many cases there is a "hot spot" on its surface, from which radiation is emitted. And since the star spins rapidly on its axis, this beam of radiation sweeps regularly across the sky like a lighthouse beacon (figure 9.4). In such favorable cases, the neutron star is identified from the startling regularity with which the pulses of radiation are observed; the star is seen as a *pulsar*. What makes the pulsar PSR 1913 + 16 discovered by Taylor and Hulse so special is that it has a companion, another neutron star, and the two orbit one another as binary stars do, in very close proximity. This system provides a near-ideal test-bed for general relativity. Hulse and Taylor and their collaborators studied it for nearly twenty years, over which time the gradual changes they measured confirmed Einstein's theory with ever-increasing precision.

The first feature of the binary pulsar is shared with pulsars in general; it is an astoundingly accurate "clock," as accurate as the best atomic clocks we can make. The "ticks" of this clock are provided by the regular pulses of radiation we detect as the pulsar spins on its axis, bringing the "hot spot" into view once every revolution. For the binary pulsar this happens every 59

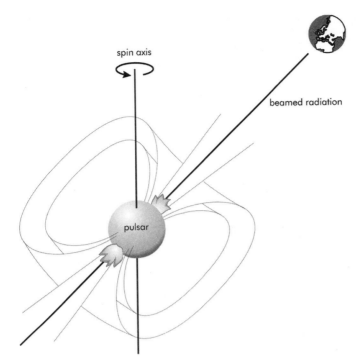

spin axis

beamed radiation

pulsar

Fig. 9.4. A celestial lighthouse. A "hot spot" on the surface of a rotating neutron star emits radiation that we detect as regular pulses. Such stars are called pulsars. The diagram illustrates how this occurs. Just as the Earth spins on its axis, so does the star. And just as the Earth has a magnetic field, with the magnetic axis aligned at an angle to the spin axis, so it is also for the star. The magnetic poles of the star are the hot spots, from which intense radiation is emitted, and as the star spins, the beams of radiation sweep around like light from a lighthouse. If each time the star makes a complete revolution the beam comes towards us, we see pulses of radiation.

milliseconds, with a regularity accurate to one part in 10^{14}! To be sure, it is gradually slowing down, again in a regular, well-measured, and well-understood fashion—but only at a rate of 1 second in 100 million years. The regularity of the ticks is modulated by the Doppler effect as the pulsar swings around its companion, and from this modulation Taylor and his colleagues have deduced with accuracy better than one part in a million the parameters which characterize its Keplerian orbit. In particular they have been able to observe the precession of its periastron (the analogue for a star of a planet's perihelion), which because the "year" for the binary (the time it takes to make a complete orbit) is only 7.75 hours (!), and because the closeness of the stars to one another (their separation varies between 1.1 and 4.8 times the radius of the sun) means that they experience very strong gravitational fields, the precession rate is some 30,000 times faster than that of Mercury. The gravitational redshift and the Shapiro time delay have also been measured, and again agree spectacularly well with the prediction of general relativity theory. But there's more. Just as an accelerated electric charge generates electromagnetic radiation, so general relativity predicts that masses in motion can generate *gravitational radiation*. In consequence, the binary pulsar should lose energy as it emits gravitational waves, and so its orbit will slowly change. This gradual change has indeed been observed, precisely as predicted, and gives the first direct confirmation of the existence of gravitational radiation.

As we have seen, the gravitational interaction is many orders of magnitude weaker than, for example, the electromagnetic interaction, making it enormously more difficult to detect gravitational radiation than electromagnetic radiation (for which all we need is a radio receiver—or our own eyes). You may be puzzled to hear that gravitational interactions are so very much weaker than electromagnetic interactions. After all, we are much more aware of the force of gravity—through our weight, for example—than of electromagnetic forces. But gravitational forces are always attractive, and our weight is the resultant pull of all the matter in the earth; electromagnetic force charges can be positive or negative and thus can cancel one another out. And because of the equivalence principle, all the components of a gravitational wave detector will be affected in the same way by a uniform gravitational field; only the "tidal effects" produced by a gravitational field which varies across the apparatus can be detected.

When electromagnetic radiation is detected, the primary effect is that electric charges are made to oscillate; in the same way, gravitational radiation is expected to make masses oscillate and so allow the radiation to be detected. The practical problem to be overcome in designing a detector, then, is to sense this oscillation, which because of the weakness of the interaction is expected to be exceedingly small. Small movements of a mirror can be observed using optical interferometry, and so a mirror in an interferometer has been proposed as the central mechanism of a detector for gravitational radiation. In an interferometer, a beam of light is split into two components in such a way that they vibrate in step with one another; these then follow different paths before being brought back together to interfere with one another, reinforcing or canceling one another out, so as to produce an alternation of light and dark fringes (figure 9.5). To amplify the effect, the compo-

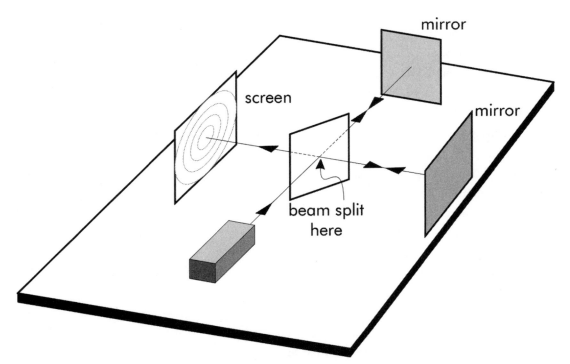

Fig. 9.5. Waves following different paths through an interferometer arrive at the detector with different phases and so lead to an alternation of light and dark "fringes." If the path difference changes, the fringes shift. In the (Michelson) version of an interferometer illustrated here, a beam of light is directed to a half-silvered mirror that acts as a "beam-splitter"; some of the light incident on the beam-splitter is reflected and some passes through. Both the transmitted and the reflected beams are aimed at mirrors that reflect them back to the beam-splitter, where they recombine and are directed toward a screen where the interference fringes can be observed.

nents of the split beam of light are bounced back and forth many times between mirrors that oscillate in response to gravitational radiation, so leading to changes in the position of the fringes. This technique has been vastly improved by the development (originally for military purposes) of extremely high-reflectivity mirrors. There are, of course, many other things besides gravitational waves which can make the mirrors shake: sound vibrations, thermal motion, earth tremors, to mention just a few. Minimizing these and amplifying the sought-for effects requires great ingenuity and very substantial resources. And to eliminate other sources of error that might mimic the

Fig. 9.6. LIGO, the Laser Interferometer Gravitational Wave Observatory. This facility houses the giant interferometer at Hanford, Washington. There is a similar interferometer at Livingston, Louisiana. (Courtesy of LIGO Laboratory)

signal of a gravitational wave, plans have been made to look for coinciding detections in detectors across an international network of three interferometers, each of kilometer-scale. This project includes LIGO (Laser Interferometer Gravitational Wave Observatory), which is being developed at two sites in the United States, Hanford in the state of Washington and Livingston in Louisiana (figure 9.6), and VIRGO (not an acronym; it is named for the constellation, which includes a possible source of gravitational waves), which is being built at Pisa in Italy. There is also a British-German collaboration (GEO 600) seeking to build an advanced version of such a detector near Hannover, which will likewise join the network. Although the cataclysmic astrophysical events that might lead to detectable gravitational radiation are rare, the success of prototype detectors that have already been constructed encourages the expectation that the planned network will be sensitive enough to start recording data early in the new millennium. An even more ambitious project, LISA (the Laser Interferometer Space Antenna), is to launch three spacecraft into orbit around the sun at the vertices of an equilateral triangle, 5 million kilometers on a side (figure 9.7)—an interferometer similar to LIGO, only 1 million times larger. The launch is at present scheduled for 2010 or sooner. Just as radio astronomy opened up a new window on the universe, so will gravitational observatories. We can look forward in this century to discoveries that will significantly extend our knowledge of the cosmos.

There is one further prediction of general relativity theory which has ex-

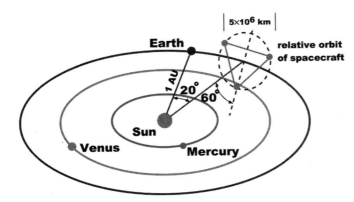

Fig. 9.7. LISA, the Laser Interferometer Space Antenna, a proposed deployment of three satellites as a triangular interferometer 5 million kilometers on a side. (Provided through the courtesy of the Jet Propulsion Laboratory, California Institute of Technology, Pasadena, California)

cited popular imagination. It also occupies a center-stage position in current research and will surely continue to do so into the new millennium. If enough matter (or energy) is compressed into a small volume, its gravitational pull is so great that nothing, not even light, is able to escape from its attraction; it has become a *black hole*. Matter may fall into it, and light may shine on it and be absorbed, but nothing can emerge from it, nothing can escape its gravitational attraction. (Well, almost nothing. There are believed to be quantum effects which modify this prediction of Einstein's classical theory.) Laplace had already speculated that something of this kind might be possible. He knew that to escape from the gravitational pull of the earth, for example, by firing a rocket upward into space, requires a launch velocity in excess of 11.2 kilometers a second. This escape velocity depends on the ratio of the mass of the earth to its radius; if the mass were greater or the radius smaller, it would be increased. For a sufficiently large mass or small radius, the escape velocity becomes as large as the speed of light, and so, argued Laplace, light could not shine out from such a body. Although Laplace's premises were Newtonian, the conclusion he came to essentially matches that of general relativity. So if a star collapses so catastrophically that it does not halt to become a white-dwarf, or even a neutron star, then presumably nothing can prevent a continuing collapse until eventually all the mass of the star is concentrated inside so small a radius that it has become a black hole. Surrounding a black hole there is in effect a surface into which matter or light can fall but out of which nothing can emerge. What happens on the other side of this "event horizon" is forever concealed from the view of any observer outside; and any one so misguided or unfortunate as to sail across this horizon will be torn to pieces by gravitational tidal forces as the mounting and fatal attraction to the center tugs harder on one part of his body than on another.

For all that we cannot expect to *see* black holes directly, their existence may be inferred from the gravitational pull they still exert on matter outside, and in particular on matter swirling around the black hole as it falls inwards like water going down a plug-hole (plate 17). The radiation from such accretion discs has been detected, and from this, together with observations suggesting that the disc is part of a binary system—that it has, so to speak, a dark twin—we can conclude that certain sources of x-ray radiation in the sky are indeed associated with black holes. There are also compelling reasons to believe that a black hole with a mass equal to that of millions of stars exists at the center of some galaxies—perhaps of most galaxies, perhaps of our own. The center of our galaxy, the Milky Way, is obscured from direct astronomical observation by intervening clouds of dust. Our armory of new astronomical techniques has penetrated those clouds, and revealed that there is indeed a black hole at the heart of that darkness. The radio source called Sagittarius A* has been observed also to emit x rays, and detailed analysis has concluded that their source is a supermassive black hole, 2.6 million times more massive than the sun, there at the center of our galaxy.

10

STRINGS

★ ★ ★

A Theory of Everything?

THE TWENTIETH CENTURY HAS SEEN PROFOUND CHANGES IN THE physicist's conception of matter. The "artificial, hypothetical atoms and molecules" criticized by Mach have been demonstrated to be as real as stars and galaxies. We didn't just dream them up. We can take pictures of them, and study their behavior. They have been broken apart and anatomized, revealing the symmetries and underlying simplicity of the standard model. Our concepts of space and time have also undergone a radical metamorphosis, emerging as the "geometrodynamics" of general relativity. Quantum mechanics has replaced the certainties of classical mechanics with a framework which we all are pleased to use, even if none of us comprehends it. Countless open questions remain. We are ever seeking, for example, to develop novel materials for electronics or other technologies. More abstract problems, too, though seemingly remote from everyday applications, urgently call for solution, attracting the attention of some physicists and exciting the wonder of nonscientists as well. At the intersection of the latest revolutionary ideas in

136

physics, not only those mentioned above but also those deriving from cosmology and the desire to find a comprehensive but essentially simple theoretical basis for all of physics, there has emerged a theory which is still in an unresolved ferment as the new century dawns. It has been touted as a "Theory of Everything," but not by its originators nor by those most closely involved in its development. It is called string theory.

The fundamental particles of the standard model still are conceived as essentially pointlike entities, even when viewed as the quanta of fields.[1] The histories that figure in the Feynman picture represent them as propagating in spacetime, tracing out lines that intersect at the points where the fields interact. Indeed, this *locality* of the interactions is one of the key building blocks for the construction of the consistent relativistic field theories on which the model depends. There is nothing akin to action at a distance. General relativity stands apart from the standard model, incorporating the forces of gravitation that are absent from the model but which must be incorporated into any putative overarching "Theory of Everything." General relativity as a classical theory is a field theory with some similarity to Maxwell's electrodynamics; the field variables are the quantities that characterize the geometry of spacetime, quantities which themselves have a dynamical behavior— hence the term "geometrodynamics," coined by John Wheeler. Wheeler gave a characteristically pithy summary of general relativity: "Spacetime tells matter how to move. Matter tells spacetime how to curve." But before string theory was developed, no one had come up with a viable solution to the problem of making general relativity compatible with quantum mechanics.

String theory is based on the notion that the fundamental entities are extended stringlike objects, rather than the pointlike particles of older theories. Just as a point particle traces out a path as it moves in spacetime, its world-line, so a string sweeps out a surface, its *world-sheet*. The dynamics of a relativistic particle in free fall can be described by requiring the path swept out to be extremal (maximum or minimum) in length. By analogy, the dynamics for the string is based on the requirement that the area of the surface swept out is extremal. There is, of course, a great deal more to it than that; the quantum fields that are defined on the world-sheet enter into the dynamics too. It turns out that the mathematics imposes very stringent self-consistency requirements, which in effect define the theory almost uniquely. The remarkable consequence is that from this rather simple starting point it seems possible to recover a consistent quantum theory of gravity coupled to other fields, which had until now eluded formulation. This quantum gravity exhibits much of what is known from the standard model of high-energy physics. Furthermore, the theory has a rich mathematical content, leading to applications in pure mathematics, especially in geometry and topology.

String theory was not originally developed to provide a consistent quantum theory of gravity; rather it first functioned as a phenomenological model to illustrate some of the properties of mesons derived from the way quarks were bound together. This was based on an idea similar to that of Faraday, who imagined "lines of force" pulling particles with opposite electric charges toward one another. So it was imagined that the quarks inside a meson were in effect held together by a sort of elastic string in tension, and the properties needed to make such a model consistent with special relativity and with quantum mechanics were worked out. This model has some very nice features: in particular, it predicts families of mesons with masses related to their spin, in the same way as experiments had found. The tension of the string, which determines the characteristic mass scale, is essentially the only adjustable constant in the model, and in order to get agreement with the observed meson masses, this had to be close to the mass of the proton—perfectly reasonable for a theory of the strong interactions. On the other hand, the new theory also predicted a state with zero mass and spin-2, and no such meson state exists. But there *is* a place in physics for a zero mass, spin-2 particle. The graviton, the quantum of the gravitational field in any quantum theory of gravity, would be just such a particle; it would have zero mass, because the gravitational waves of general relativity, just like electromagnetic waves, propagate at the speed of light, and spin-2, because that is what a tensor theory requires it to have. Could it be that the model did not apply to meson theory, but indicated how to construct a quantum theory of gravity? This possibility accounts for the extraordinary surge of interest in string theory since the 1980s.

Before string theory could become the foundation for a quantum theory of gravity a number of hurdles had to be cleared. The easy one was changing the mass scale, which could no longer be the one appropriate to hadron physics, the scale of mesons and protons mass. Instead, it has to be the scale set by Newton's gravitational constant, which together with the speed of light and the Planck constant determines unique values for scales of mass, length, and time. These are the so-called Planck mass (approximately 10^{19} times the proton mass), the Planck length (about 2×10^{-35} meters), and the Planck time (approximately 5×10^{-44} seconds). These are remote, by many orders of magnitude, from the scale of phenomena in any experiment we are able to devise, but they set the scale at which gravitational effects have to be included when considering quantum phenomena. And they also characterize the cosmic regime in the immediate aftermath of the big bang of creation.

The next problem to be overcome was one which had plagued the string model even when it was being used to describe mesons: the theory seemed to make consistent sense only if spacetime has more than four dimensions! Specifically, for a theory with only bosons twenty-six dimensions are required, while one which also includes fermions can be accommodated in ten dimensions. It is hard to accept more than four dimensions in a theory with a length scale characteristic of the strong interactions, but is not so outrageous a proposal for a theory of gravity. As long ago as the 1920s it had been suggested by Theodor Kaluza and Oskar Klein that there might be an additional, fifth dimension which was somehow restricted to a very small scale, curled up like a small circle rather than extending like the other four to infinity. This Kaluza-Klein idea incorporated a hint at the unification between electrodynamics and general relativity, a mirage pursued by Einstein through much of his later years. Perhaps the world really has ten dimensions, with all but four "compactified" to the tiny scale of the Planck length. But even with this suggestion taken on board, string theory did not sail smoothly.

The theory as it stood was unable to avoid the conclusion that there existed particles which travel faster than light—*tachyons* (from the Greek word for "swift")—and such a hypothesis opened up more difficulties than the theory might otherwise solve. What rescued string theory was the incorporation of *supersymmetry*. As a mathematical symmetry relating bosons to fermions, this had first been suggested in 1971, in a paper by Yuri Golfand and

Evgeny Likhtman. (Golfand was shortly to become a *refusenik*, having had his application to emigrate to Israel rejected, with consequent penalties from the Soviet authorities.) Supergravity theories soon followed, in which the symmetries of spacetime were extended in the way suggested by this supersymmetry, and supersymmetric models of particle physics also appeared, predicting boson partners for the known fermions (squarks as partners for quarks, etc.) and fermion partners for bosons (e.g., gluinos and photinos partnering gluons and photons respectively). These exotic mates have not (yet!) been discovered, but they are sought for in many of the experiments currently underway. And from what we observe at the energies already studied, it looks as though such particles really do play a part in processes at higher energies yet.

Turning strings into superstrings eliminated the tachyons, but there was still another obstacle to be overcome: the *anomalies*, that arise when the theory is made into a quantum theory. We have already discussed the crucial role of gauge symmetries in removing the infinities from field theory by making the theory renormalizable. Emmy Noether, one of the outstanding mathematicians of the century, developed a celebrated theorem that relates the existence of a gauge symmetry to a current, which is conserved. (Noether, despite the prejudice against women in European universities, was appointed to the faculty at Göttingen but lost her job there in 1933 under pressure from the Nazis. She fled to the United States where she taught at Bryn Mawr and at the Institute for Advanced Study in Princeton.) The familiar electric current furnishes an example of current conservation: although electric charge can move about—and a current is nothing other than a flow of charge—the total electric charge in an isolated system never changes. Noether's theorem holds for classical field theory. But sometimes in the corresponding quantum field theory, notwithstanding the presence of a gauge symmetry, the Noether current is not conserved. In that case we have an anomaly. And although we have been talking about string theory, not quantum field theory, from the point of view of the *world-sheet*, string theory *is* a quantum field theory, with fields living in the one-space, one-time, dimensioned surface which is the world-sheet. This $1 + 1$ dimensional field theory has to be free of anomalies for the string theory to be consistent.

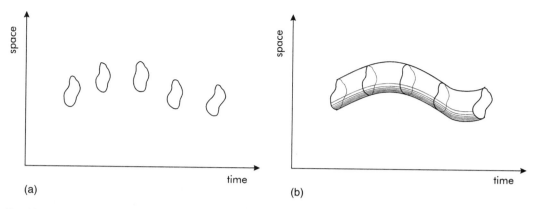

Fig. 10.1. The world-tube of a string. (a) "Stroboscopic" view of a string at successive times. (b) "Time-lapse" picture of the same string, showing the world-tube traced out in spacetime.

In a remarkable tour de force, Michael Green and John Schwarz showed that these anomalies cancel out when the string is provided with additional symmetries—supersymmetries—in a way which for self-consistency is almost uniquely determined. And so was set in motion the superstring band wagon, which continues to roll and roll. From the work of Green and Schwarz and those who followed them, it emerged that there were just five consistent, anomaly-free superstring theories. And any such theory, it turns out, will inevitably contain a theory of gravity, which at low energies (and here "low" energy still encompasses a range up to around a thousand billion times the energy achieved in any existing or proposed particle accelerator) is identical with Einstein's general relativity. Put another way, it is only at distances a million billion times smaller than the size of an atomic nucleus that "stringy" departures from Einstein's general relativity will be manifest. So string theory can be said to *predict* Einstein's theory of gravity. You might prefer to say retrodict, but remember that string theory was not devised with the objective of confirming general relativity. And it does much more than that.

Think first of a loop of string moving through space (figure 10.1). Its world-surface is a sort of tube through spacetime. The characteristic length scale—related to the tension of the string, the one free parameter in the theory—is the Planck length. This is so tiny that even on the scale of particle physics the tube is so narrow as to resemble a line—just like the world-line of an elementary particle. But there are fields "living in" the string's world-sheet, and the string itself can vibrate like a stretched rubber band, with an unending sequence of different modes of vibration, at ever higher frequencies. Each of the different modes of vibration has its characteristic energy, and behaves like a different kind of particle, which means that this one loop of string can behave like many different kinds of particle, depending on its internal vibrations. So the world-line, which this narrow tube resembles as it threads through spacetime, could be that of any one of many different possible particles. The wonderful thing is that among these possibilities we find not only the zero-mass graviton, with all the properties needed to make it fit Einstein's general relativity, but also all the fundamental particles of the standard model. But there are still more unexpected positive features of superstring theory.

General relativity, when treated as a "conventional" quantum field theory, has all the infinities characteristic of a nonrenormalizable theory because quanta in such theories are essentially *pointlike* excitations of the field, and their interactions are *localized* as isolated events in spacetime. This localization of interactions is an essential ingredient of relativistic quantum field theory, but it is also the source of the infinities which plague such theories, unless they can be tamed by renormalization. And for general relativity, Einstein's theory of gravity, it turns out that this is not possible. But the *string* theory of gravity is not a quantum field theory of the conventional kind; the tube traced out by a string in spacetime only resembles a line when its diameter is too small to be of significance. But whereas there is a singular *point* where a world-line branches into two, like a Y, if the arms of the Y were in fact tubes, there is no such singularity. And from this it follows that the string-theory analogues to Feynman diagrams contain none of the singularities that in field theory lead to infinities. So the quantum theory of gravity that emerges from string theory is free from infinities; string theory leads to a consistent quantum theory of gravity, something that eluded physicists for at least half a century. To see how this comes about, one must shrink the tube till it resembles a line, and then it becomes apparent that Einstein's theory has to be modified, but in a very precisely determined way. The corrections introduced by these modifications are quite negligible when one is concerned with the gravitational phenomena so far observed, because they are quantum effects that are far too small to be of consequence for such large-scale phenomena.

Superstring theory, therefore, promises to unify the standard model with a quantum theory of gravity, without awkward infinities and without arbitrary constants to be adjusted once the string tension has been chosen. I have to say "promises," because a great deal remains to be done before these results can be secured. For one thing, as already mentioned, consistency demands that the spacetime in which the string moves has to have *ten* dimensions, since we insist, of course, on a superstring which has both fermion and boson excitations. So why do we seem to have just four, three of space and one of time? Most likely this happens in the manner proposed originally in 1919 by Kaluza in a letter to Einstein, and later extended by Klein. Six of the original ten dimensions become "spontaneously compactified," which means that instead of extending indefinitely like a straight line, these compactified dimensions are each more like a tiny circle, with a circumference of the order of the Planck length. They are so tiny that we are unaware of them—but not of their consequences. It turns out that there are many different ways that this compactification can happen, so what started off as a rigidly restricted set of possibilities—just five consistent superstring theories—now seems once more to have opened up into an embarrassingly large range of choices. Theorists are inclined to believe that a satisfactory fundamental theory ought not to be so arbitrary, to involve so much free choice, so many opportunities to tweak the theory to make it fit the facts. This prejudice is essentially aesthetic, but it is supported by the success story of the standard model and the whole development of high-energy particle physics and quantum field theory, on which the model is based. There is not much scope for arbitrary tweaking with the standard model!

More recently, in what has been described as the second string revolution, a whole new avenue for future exploration has opened up, along which there have already been tantalizing glimpses of the way that the "rigidity" in the theory may be restored.[2] It has been established that superstring theories possess additional symmetries, called *duality* symmetries. We find a simple kind of duality symmetry in Maxwell's electrodynamics, and it was long ago

Fig. 10.2. *Sky and Water* by M. C. Escher. There is a kind of duality between the birds and fish. (© 2001 Cordon Art-Baarn-Holland. All rights reserved.)

explored by Dirac; that is, the basic equations remain unaltered if the electric and magnetic fields are interchanged. Duality may be thought of as something akin to the symmetry in some of the etchings by M. C. Escher (figure 10.2). In string theory we find a duality like that in the Maxwell theory and also another kind, one in which, for example, a solution of the equations with a dimension compactified to become a small circle is essentially equivalent to one with a large radius.[3] Other dualities and "mirror symmetries" relate solutions to the equations derived from one of the five consistent superstring theories, compactified in a certain way, to solutions from another, compactified in a different way. It is conjectured, with growing confidence, that all five superstring theories and their various compactifications are related to one another and are just different realizations of a still more fundamental theory, which one of its founders and leading advocates, Ed Witten, has called M-theory. He said, "M stands for 'Magical,' 'Mystery' or 'Membrane,' according to taste." Others have suggested that M stands for 'Mother.' M-theory lives in eleven dimensions, not ten. We do not know how to formulate this hazily perceived theory, but only that in different extremes it yields the previously established consistent string theories. I myself believe that this idea, which as I write is one of the "hottest" topics in high-energy theory, is not just a nine days' wonder but will mature to shed deeper insights into the structure of space, time, and matter.

Such speculations have also supported another slate of ideas, some of which relate to topics touched upon in earlier chapters. Let us return for a moment to "ordinary" quantum field theory, in which the fields obey equations of motion very much like the classical equations that describe more familiar waves, say, the ripples on the surface of a placid canal. In quantum field theory, these waves are quantized to become what we may call the elementary excitations of the field and are identified with the particles associated with the field (photons for the electromagnetic field, for example). Because the fields interact, these excitations also interact, and in this fashion particles scatter from one another, transmute from one kind to another, are created or destroyed. Put in another way, the result of the interactions is that the equations of motion are no longer linear. And it also happens that just as particles with different energies travel at different speeds, so also do waves with different frequencies, a phenomenon called *dispersion*. Now when a wave equation exhibits both nonlinearity and dispersion, it can support a completely different form of excitation. The historical discovery of this phenomenon in 1834 by John Scott Russell, as he watched a barge towed along the Edinburgh-Glasgow canal, is so charmingly told as to be worth repeating here:

I happened to be engaged in observing the motion of a vessel at a high velocity, when it was suddenly stopped, and a violent and tumultuous agitation among the little undulations which the vessel had formed around it attracted my notice. The water in various masses was observed gathering in a heap of a well-defined form around the centre of the length of the vessel. This accumulated mass, rising at last to a pointed crest, began to rush forward with considerable velocity towards the prow of the boat, and then passed away before it altogether, and, retaining its form, appeared to roll forward alone along the surface of the quiescent fluid, a large, solitary, progressive wave. I immediately left the vessel, and attempted to follow this wave on foot, but finding its motion too rapid, I got instantly on horseback and overtook it in a few minutes, when I found it pursuing its solitary path with a uniform velocity along the surface of the fluid. And having followed it for more than a mile, I found it subside gradually, until at length it was lost among the windings of the channel.

Another name for a solitary wave of this kind is a *soliton*, and attention has been paid to the suggestion (for example, by Tony Skyrme) that some at least of the particles we observe are more like solitons than elementary field excitations.

There *are* solitonlike objects to be found in superstring theory, and the wonderful thing is that the elementary excitations in one kind of superstring can be related by a duality transformation to a soliton excitation in another superstring formulation. This interchange of solitons and elementary excitations by a duality transformation allows one to use the techniques of perturbation theory to explore otherwise inaccessible aspects of the theory where the interactions are strong. This has a precise analogue in statistical mechanics, where a similar duality relates properties at high temperature to properties at low temperature.

And there are also solitonlike solutions to the equations of general relativity. One of these was found as long ago as 1916 by Karl Schwarzschild; it is the famous *black hole* solution to Einstein's equations, and its importance was immediately recognized. The object it describes has spherical symmetry and does not change in time; only one parameter enters into the solution, which can be identified with its mass. Since the solution is valid where there is no matter present, it can be used to describe the curvature of spacetime *outside* a spherically symmetric, static body—and for most purposes we may regard the sun as such. So the first application of the Schwarzschild solution was to derive the consequences of general relativity for the motion of the planets and other bodies within our own solar system, with the spectacular successes already described. The Schwarzschild geometry does not apply *inside* the sun, where matter is present, but does join smoothly to a solution appropriate there rather than to the "empty space" around it. This is just as well, since for every star there is a critical radius related to its mass, and if all the mass is concentrated within that radius, so that the solution continues to be valid right down to within the critical radius, we have what is truly a black hole. You will be relieved to hear that the sun is in no danger of this catastrophe: its critical radius is about 3 kilometers, as against its actual radius, about 700,000 kilometers. So for the sun to become a black hole, all the matter in it would have to be squeezed into a sphere of radius 3 kilometers—which is not likely to happen!

But as we have seen, it *is* possible for a black hole to be formed from the collapse of a sufficiently massive star under its own weight, something that is predicted to occur after it has exhausted all of the thermonuclear reactions that previously buoyed it up. Several astronomical objects are believed to be just such black-hole stellar remnants. Other candidates for black holes are the cores of quasars and other so-called "active galaxies," since apart from the gravitational energy that would be released by whole stars falling into a supermassive black hole (with a mass believed to be several million times that of the sun), we know of no source of energy equivalent to what active galaxies emit.

In classical, nonquantum general relativity a black hole, once produced, can only increase in mass as it draws into itself matter or radiation from its surroundings. If nothing were to fall into it, such a black hole would persist, growing in mass but otherwise essentially unaltered in time. Observed (safely!) from a distance, it would make its presence known only from its gravitational attraction, which is determined by its mass. Quantum effects change all that, however, and in a most surprising way. In 1972 Jacob Bekenstein demonstrated intriguing analogies between the properties of black holes and the laws of thermodynamics. Two years later, these were clarified by Stephen Hawking (figure 10.3), who discovered that quantum effects make a

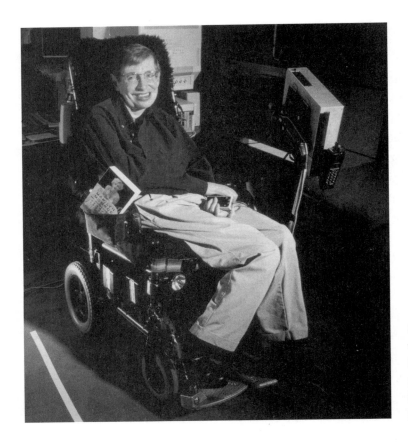

Fig. 10.3. Stephen Hawking, who today holds the Lucasian Chair in Cambridge previously held by Isaac Newton. (© Kimberly Butler Photography)

black hole *radiate*—just like a black body, as studied by Planck—and at a temperature that is inversely proportional to its mass. So because of quantum effects, an isolated black hole will lose energy. But remembering that energy and mass are interchangeable ($E = mc^2$ again!), loss of energy means loss of mass; the black hole loses mass, and since its effective temperature is *inversely* proportional to its mass, its temperature *increases*, and so it radiates faster and faster. In deriving this result, Hawking was unable to take account of the "back-reaction" on the black hole of the radiation emission, so that no one is sure what the very last stage of this quantum radiation process would look like, when the energy remaining is of the order of the Planck energy—about 1.8×10^9 joule, or enough energy to keep a 60W light bulb going for a year. But it is definitely predicted that in the final tenth of a second there would be released about as much energy as from an H-bomb explosion. But the military, monitoring for possible violations of test-ban treaties, report that no such violent explosions have been observed in the vicinity of Earth. This is in fact not so surprising, since a stellar-mass black hole would have a temperature only a millionth of a degree above absolute zero, and its Hawking radiation loss would be completely negligible compared with the rate at which it would absorb energy from the 2.7 K cosmic background radiation.

Although there is strong evidence in favor of stellar-mass black holes and also of supermassive black holes, the possibility of "mini" black holes is still open. They are one way to account for some of the "missing mass" in the universe, and some cosmologists like this idea, because it brings consistency to their preferred model for cosmic evolution. Mini–black holes might have been produced in the big bang of creation, and it is not an idle conjecture to suppose that they do occur. But my own guess is that they do not occur as astronomical objects.

Classical Einsteinian general relativity allows other black-hole solutions besides the Schwarzschild one. In some, the black hole can carry electric charge, or spin like a top. Since superstring theory contains general relativity, you will not be surprised to learn that it, too, has black-hole soliton solutions with mass, charge, and spin. For some special values of these quantities, it turns out that these predicted states do not have any Hawking radiation, and so even in quantum theory they are *stable*. This lends some support to the suggestion made by Abdus Salam and others that at least some of the elementary particles are in fact black holes!

Some of the solutions to the equations of M-theory describe objects that are one-dimensional objects in space—strings. There are also solutions that instead describe two-dimensional surfaces, something like membranes. And there are also objects with even more dimensions; these, for want of a better word, are all called *branes*. It has been suggested that our familiar three-dimensional world may in fact be a 3-brane living in the higher dimensions of M-theory. It may be that we can only become aware of what lies outside our 3-brane world via gravitational effects that bear the trace of what lies elsewhere. To be sure, this may sound like a fantastic invention, unrelated to the real physical world. But the equations of M-theory do have solutions of this kind, and there seems to be no reason to reject them.

The duality that implies that curling up a space dimension into a circle of radius R is equivalent to curling it up into a circle of radius proportional to $1/R$ suggests that on the scale of the Planck length our notions of space need radical revision. It seems to imply that a distance between two points of, say, half the Planck length, is just the same as a distance of twice the Planck length—so it doesn't make sense to talk of separations on this scale. Pictured another way, what seems on our gross scale to be smooth and featureless space becomes at this level of refinement what Hawking has called "space-time foam": quantum fluctuations generating evanescent black holes, which form and reform.

The equations of general relativity have solutions which describe a kind of reverse counterpart to black holes—*white holes*. If we think of a black hole as a sort of funnel in space, which like a tornado sucks matter into itself, then a white hole is an inverted funnel from which matter pours out. The funnel of a black hole may be connected by a "wormhole" to the throat of a white hole, and the matter sucked into the black hole may then be ejected from the white hole. And the white hole may be far away from the black hole to which it is connected! This sounds more like science fiction than science prediction—or perhaps like *Somewhere over the Rainbow*—but the possibility is at least not inconsistent with our present understanding of general relativity, and also with what M-theory seems to permit.

In the next chapter I will try to give an account of our present understanding of cosmology, a wonderful interplay between general relativity and particle physics, the physics of the very large and of the very small. The theory of superstrings has not yet answered some of the open questions (of detail, rather than of principle) in big bang cosmology, but I believe it will. In any case I find it awesome that there are such intimate links between physics at the smallest scales, set by the Planck time and the Planck length, and the physics of the cosmos as a whole, with its time scale of the order of 15 billion years and its length scale of as many light-years. The fundamental principles that link these two scales—separated by some sixty orders of magnitude—have only begun to emerge in the late twentieth century and will surely continue to enrich physics in the twenty-first.[4]

11

IN THE BEGINNING

★ ★ ★

The Big Bang—and After

THE MYSTERY OF THE COSMOS, ITS ORIGIN AND EVENTUAL FATE, has aroused wonder and conjecture since time immemorial. The majestic passage of the stars across the night sky inspires awe. In California I was once asked by a student taking a course on astrophysics why it was that he could not just lie out on the hillside, gaze up at the stars, and "make the connection between the microcosm and the macrocosm." He did not know how close he was to a very profound truth! For to understand the details of the "big bang" cosmology, now so convincingly supported by data, we must know something about particle physics and what happens at the scale set by quantum gravity, that is, the Planck scale, with a characteristic length of just 10^{-35} meters. But let us start nearer to home.

On a clear day, you can feel the heat and see the light radiated from the sun. Although we are 1.5×10^{11} meters away, the intensity of radiated energy received at this distance from the sun is about 1.5 kilowatts on every square meter of Earth's surface. This prodigious outpouring of energy has lasted for some billions of years. Its source, as with all stars, is thermonuclear—principally, the conversion of hydrogen to helium through fusion. Hans Bethe (one of the grand old men of physics, still active today, in his nineties) largely worked out the subtle and complex details of these reactions in 1938. Eventually the sun's hydrogen will be used up, and then other reactions will take place, producing oxygen and carbon as the helium is itself consumed. When all their fuel is exhausted, old stars can collapse to form white dwarfs, or they can explode violently as supernovae. A supernova explosion really starts as an implosion: the star collapses under its own weight, compressing its core to initiate a dramatic last gasp of thermonuclear transformation. Stars live a long time, so supernova explosions are rare, only one or two in a hundred years in a typical galaxy. Matter ejected from a supernova explosion may then be drawn by gravity to mix with gas clouds where a new generation of stars is being born. Our planet, Earth, was formed from the same condensing cloud of gas that gave birth to the sun (a third-generation star), so to ask where we came from we need to know where the sun came from.

ALMOST ALL OF THE CHEMICAL ELEMENTS IN THE UNIVERSE, APART from hydrogen, helium, and lithium, originated in the thermonuclear reactions in stars and were dispersed by supernova explosions. As noted before, we are made from stardust. The iron that reddens our blood resulted from the decay of radioactive nickel produced in a supernova. From the spectrum of sunlight we know that the gas and dust that condensed to form the sun contained such elements, which could only have originated from the supernova death of stars from an earlier generation. But the hydrogen in the cosmos, which is still its most abundant element, and most of the helium are older than any of the stars. Less than a microsecond after the big bang, when the temperature of the expanding universe had fallen sufficiently, quarks could join together to make protons, which would, 300,000 years later, combine with electrons to make atomic hydrogen.

We now have a wonderfully robust understanding of how "ordinary" matter was formed after the big bang. Piecing together the sequence of changes brought about by the expansion of the universe and its resultant cooling is one of the supreme achievements of cosmology. It rests on a few key observations, and a great deal of detailed physics: high-energy particle physics, statistical physics, nuclear physics, and general relativity. To set the stage, we must turn to the last of these. Einstein's theory of general relativity relates the structure of spacetime, its geometry and dynamics, to the presence and distribution of matter. And as John Wheeler put it, "matter gets its moving order from geometry." Einstein's equations, difficult though they are in general, become quite tractable when applied to cosmology. This is because of the assumption that space and the distribution of matter in the universe are homogeneous, the clumping up of matter to form galaxies and stars being irrelevant to the large-scale structure.

In 1922 Aleksandr Friedmann had already shown that with this assumption Einstein's equations had a solution that corresponded to an expanding universe. (This was before Hubble's observations had led to the same conclusion.) His paper in *Zeitschrift für Physik* received a critical response a few weeks later from Einstein, who wrote: "The results concerning the nonstationary world, contained in [Friedmann's] work, appear to me suspicious. In reality it turns out that the solution given in it does not satisfy the field equations." Friedmann wrote to Einstein, sending him details of his calculations, asking: "Should you find the calculations presented in my letter correct, please be so kind as to inform the editors of the *Zeitschrift für Physik* about it; perhaps in this case you will publish a correction to your statement or provide an opportunity for a portion of this letter to be published." Einstein did not receive the letter for some months—he was on a trip to Japan—but when he did read it, he wrote to the journal to say that his criticism was "based on an error in my calculations. I consider that Mr Friedmann's results are correct and shed new light." But Einstein considered Friedmann's results to be of mathematical interest only. Though unaware of Friedmann's paper, the Belgian astronomer—and priest—Georges Lemaître published a note in 1927 that demonstrated the same result. Neither Friedmann's nor Lemaître's work received much attention until 1931, when Eddington published Lemaître's note in English translation, and Lemaître introduced his ideas on "the primeval atom" in lectures given in London at the British Association for the Advancement of Science. This could be said to mark the birth of the

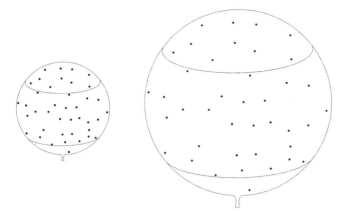

Fig. 11.1. A model for the expanding universe, the surface of a balloon. The spots represent galaxies, and as the balloon expands, they move farther apart from one another.

big bang cosmology, which is at the heart of our present understanding of the birth of the universe.

Along the same lines, Howard Robertson and Arthur Walker also found general cosmological solutions to Einstein's equations, and the Friedmann-Lemaître-Robertson-Walker solutions provide the basis for our present models of cosmology. The solutions of interest all posit a "big bang" origin for the universe; that is, they describe a universe that expanded (and continues to expand) from a mathematical singularity, when all of space was concentrated in a single point.

To picture how the universe grew from some initiating cataclysmic explosion, imagine a universe with just two space dimensions, of finite size but without any boundary, like the surface of a balloon. As the balloon is blown up, its surface expands: the universe gets bigger. If the balloon has spots on it, which represent galaxies, as time advances and the balloon gets bigger, the spots on its surface, like the galaxies in our universe, get carried away from one another (figure 11.1). The big bang for this model universe, then, is the moment when the balloon was so small as to be considered a point. It is perhaps not easy to step up to three dimensions, but that is what you must try to do in order to envisage the conventional big bang picture of a finite space without boundary expanding as time passes.

The Friedmann-Lemaître-Robertson-Walker solutions to Einstein's equations allow us to extrapolate backward from the present state of the universe to its origin in a stupendous explosion. The equations relate the size of the observable universe and the density of matter and energy it contains to its age at any epoch, and statistical thermodynamics relates the energy and matter density to the temperature. To be sure, the assumptions of the model are probably not to be trusted for the *very* first instants after the big bang, but from about 10^{-32} seconds onward they are increasingly reliable. The expansion initiated by the big bang has led to a fall in temperature from something like 10^{25} K at 10^{-32} seconds to the 2.7 K of the cosmic background radiation today. The whole of the presently observable universe was, at 10^{-32} seconds, compressed to less than a meter in size. At such density and temperature, the behavior of matter is believed to be surprisingly simple. Particles and antiparticles of all kinds incessantly interacted, combining in and reemerging from the fierce radiation that filled all space. Despite this frenzied activity, both matter and radiation were near to being in thermal equilibrium, and so obeyed the laws of statistical mechanics appropriate to these conditions, which means that only the prevailing temperature is needed to calculate their salient properties. And the temperature in turn can be deduced by working backward from what we observe today. What makes this cosmological model so convincing is that from a few plausible assumptions there follow a wealth of predictions that can be checked against observation.

At 10^{25} K the characteristic energy of collisions between particles is so high that, according to the standard model of particle physics, the electromagnetic and weak nuclear forces would be unified as in the electroweak theory before spontaneous symmetry breaking (see chapter 8). Matter was then in the form of quarks and leptons and their antiparticles. By about 10^{-12} seconds after the big bang, the temperature would have fallen sufficiently for the electromagnetic and weak interactions to separate, and after about a microsecond the quarks would be able to combine to form nucleons—protons and neutrons. But it was still far too hot for stable atoms or even nuclei to exist, and matter would have been in the form of a dense plasma of nucleons and leptons, bathed in electromagnetic radiation—all at a temperature of a million million degrees, 10^{12} K. (Plasma is a state of matter that differs from a gas in that it contains independent positive and negative electrically charged particles, rather than electrically neutral atoms and molecules. This state of matter is not rare: 99 percent of known matter in the universe is in a plasma.) Protons, neutrons, and electrons, and their antiparticles, coexisted in approximately equal numbers; photons and each of the different kinds of neutrinos were some 10^9 times more abundant.

By the time the universe was a second old, the equilibrium between protons and neutrons that had prevailed until then could yield to the mass difference between them. A neutron is slightly more massive than a proton, and a free neutron can decay into a proton, producing at the same time an electron and an antineutrino; the reverse interaction can also occur, but as the temperature fell it became increasingly rare. One of the crucial parameters that enters into cosmology was then fixed: the relative abundance of neutrons to protons (about 1 to 5 at that time). From then on, the neutrinos have had negligible further involvement through their nuclear interactions. But their presence may still be of cosmological significance, since they are so abundant, especially as we now believe that they have mass. This is because one of the key parameters determining the evolution of the universe is the average density of mass it contains—and a neutrino mass even as small as that suggested by the Super-Kamiokande observations would still imply a contribution comparable with that from all the visible stars.

By one minute after the big bang the temperature had fallen to 10^{10} K, and it became possible for neutrons and protons to combine to form nuclei. There is no stable combination of just two neutrons, nor of two protons. So the first step would be the binding together of neutron-proton pairs to form deuterons, the nuclei of deuterium, a heavy form of hydrogen. Deuterons are very fragile, and so are easily broken apart. It was only in the hot, dense environment of the early universe that collisions between a pair of deuterons, or between a deuteron and a proton or a neutron, could occur sufficiently often that despite the odds against successful bonding, some nuclei of helium, tritium (tritium is a still heavier, radioactive form of hydrogen), and helium-3, respectively, were produced. Only these and others of the simplest nuclei would have formed with any abundance, because collisions bringing together enough protons and neutrons to form anything as complicated even as beryllium would have been far more rare than those productive of deuterium or helium. After about ten minutes all these interactions would be over. The relative abundance of the light elements formed in those first few minutes can tell us about the temperature and density, and is another of the key pieces of data that impose such exquisite constraints on big bang cosmology. The expansion and cooling of the universe continued, with matter still in the form of a dense plasma, composed now mainly of hydrogen and helium nuclei, with a tiny trace of lithium, all in electrical balance with electrons and all still in equilibrium with the radiation, which had a black-body spectrum corresponding to the steadily falling temperature.

When the universe was 100,000 years old, the typical energy of a thermal collision had fallen below that which would break atoms apart through ionization. Now, for the first time, electrically neutral atoms could form. It would take another 200,000 years before most of the ionized atoms had combined with electrons to become neutral atoms. Matter no longer took the form of a plasma of electrically charged nuclei and electrons but was now a neutral gas of hydrogen and helium. This brought about a dramatic change in the interaction of matter with radiation, since unlike the primordial plasma, a gas of neutral atoms is essentially *transparent* to light. For the first time light could shine through, and matter and radiation became *decoupled* in their subsequent evolution. What we observe now as the cosmic background radiation is nothing other than the thermal black-body radiation which existed at that time, now cooled by the continuing expansion of the universe from the 3000 K which prevailed at 300,000 years to the 2.7 K of today. And the temperature of that radiation is another crucial piece of data that helps secure the big bang model of cosmology (figure 11.2).

What are the assumptions that support this scenario for the birth of the universe? One, as already noted, is that the equations of general relativity, which relate matter to geometry, are trustworthy. Another is that we may simplify those equations by assuming the "Copernican" cosmological principle, that the universe is, at the largest scales, homogeneous. The Friedmann-Lemaître-Robertson-Walker equation which then follows still requires some further input. This equation connects the way the size of the universe changes in the course of time to the density of its energy and the pressure of matter and radiation it contains.[1] The Planck radiation law relates the energy density and pressure of radiation to the temperature, and there is a similar formula appropriate for matter (in the form of a gas, for example).

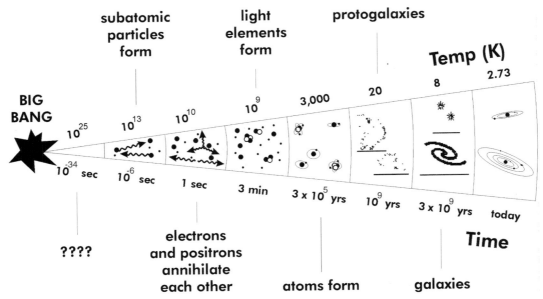

Fig. 11.2. The cooling of the universe as it ages. As the universe expanded, the wavelengths of the black-body radiation generated 300,000 years after the big bang have been stretched a thousandfold. And its temperature has cooled from around 3,000 K to 2.7 K. (With thanks to Professor M.J. Griffin)

The resulting equations describe an expanding universe, which is what we indeed observe. And one of the key observations that can be used to determine the parameters that enter the equation is the Hubble constant, which is given from the present rate at which the universe is expanding through the relation between the speed at which distant galaxies are receding and their distance. The exact value of this constant is still somewhat uncertain, because it is hard to make reliable determinations of the distance of the galaxies. But most astronomers and cosmologists are converging on an agreed value, from which the age of the universe can be estimated as around 13 billion years.

Theorists strive for simple answers, but nature may not be so kind as to provide them! Before Hubble's observations, and before Friedmann's solution to the equations of general relativity, Einstein believed that the geometry of the universe was unchanging on a cosmic scale. In searching for a static solution to his equations, he recognized that something was needed to counter the gravitational attraction of matter, which would otherwise tend to fall together, leading to a *collapsing* universe, not a static one. This led him to modify his equations by introducing a term governed by what is known as the cosmological constant, Λ (lambda). This has the effect of countering the collapse and makes it possible to find a static homogeneous solution after all. In the same year, 1917, Willem de Sitter found another solution to Einstein's equations which was also homogeneous and static but had the remarkable property of predicting a redshift proportional to distance. De Sitter's universe was empty of matter, so it was not very useful as a model of our own—but it has found surprising applications to problems in M-theory (which we met in chapter 10)! It also is an ingredient in the theory of inflation, which we will describe below. As soon as Hubble showed that the universe was *not* static, Einstein rejected the cosmological term and regretted that he had ever introduced it, because there appeared to be no need for it. (He is said to have described its introduction as his greatest blunder.) Until a few years ago, observations suggested that it was extremely small, so theoretical prejudice led to a consensus view that it was exactly zero.

The other parameter that must be entered into the Friedmann-Lemaître-Robertson-Walker equation is related to the density of mass and energy in the universe. If there is enough matter in the universe, its gravitational attraction will eventually reverse the expansion, and the universe will begin to collapse, ending finally in a "big crunch."[2] Or maybe a "big bounce!" The parameter that determines this is called omega (Ω). If omega is more than 1, there will be a big crunch, but if omega is less than or equal to 1, the universe will expand forever. The critical value 1 that separates these two fates for the universe leads to eternal expansion, but at a rate which is forever slowing down. Omega is the sum of two terms, one coming from lambda, and one from the density of mass and energy. So a crucial issue for observational cosmology is to pin down the contribution to omega from the mass density, but this turns out to be surprisingly difficult. We can estimate the mass of a galaxy by reckoning on 10^{11} stars, each with a mass comparable to that of the sun, and we can observe the density of galaxies in the universe. But we have good reasons to suppose that this approach seriously underestimates the total mass. Firstly, not all of "conventional" matter is in the form of visible stars. The tenuous gas between the stars, and even between the galaxies, contains a great deal of mass. Further evidence suggests that a galaxy may have as much as ten times the mass that can be attributed to its visible stars, the remainder coming perhaps from dark, dead stars or black holes. Some of this evidence comes from detailed observations of the rotation of galaxies, as deduced from the Doppler shifts of the stars they contain. But the most compelling reason to suppose that there is much more *dark matter* in the universe than that which we can see comes from a different source: the proportionate amounts of the light elements that were created when the universe was a few minutes old.

Most of the atomic nuclei in the universe have not changed since they condensed out from the sea of protons and neutrons in the first few minutes after the big bang. These primeval nuclei were created in thermonuclear reactions generated by the enormous temperature that brought protons and neutrons into energetic collisions. On the other hand, it was only because the temperature was falling rapidly that the nuclei were not broken apart again as soon as they had been created. Meanwhile, the abundance of free neutrons was already being eroded, since they live on average for only about seventeen minutes before decaying radioactively into protons. The ratio of different kinds of nuclei in the resulting mixture can be calculated, and it depends with extreme sensitivity on the density of matter present when these reactions commenced. The initial proportions of 1 neutron to 5 protons can be determined from known features of high-energy particle physics, and with this input, the relative amounts of deuterium, of the two different isotopes of helium, of lithium, of beryllium, etcetera, in the early universe can all be calculated. This is not easy! It involves a huge computation, first performed by Jim Peebles in 1966 and by Robert Wagoner, William Fowler, and Fred Hoyle in 1967, and repeated several times since then. When the result is compared with the present abundance of each of these elements (correcting for the effect of stellar evolution), we get a very precise determination of the density of matter at the time these primordial nuclei were formed.

The next step in the chain of argument is to relate the ancient density of matter to the present one, since we know by how much the universe has expanded in the interval. A puzzle is revealed: given the primordial ratio of the elements, the density of visible matter now appears to be too small. But as we have already noted, "dark matter" can account for the missing mass. Speculations concerning possible contributors to the missing mass include the whimsically named WIMPs, for "weakly interacting massive particles," and MACHOs, for "massive astronomical compact halo objects," possible orphan planets or "brown dwarfs," would-be stars that never became massive enough to ignite their thermonuclear fuel. These might exist in the extended halo that surrounds our galaxy, and likewise in other galactic halos. To find them, investigators are looking for the distortion their gravity would produce on the images of stars beyond them, an effect predicted by general relativity. Some theories of particle physics which go beyond the standard model predict particles called axions, and these might be candidates for WIMPs. They too are being sought, through the same kind of experiments used to detect neutrinos. The kind of dark matter to which neutrinos might contribute is called "hot dark matter," since such particles would be expected to have had speeds close to the speed of light until their interactions with other matter became so rare that they were "decoupled." WIMPs and MACHOs, on the other hand, would be examples of "cold dark matter." Hot and cold dark matter would imprint the large-scale structure of the universe in subtly different ways, but present data are not sufficiently accurate to decide between them. Most astrophysicists favor a judicious mixture of the two as the explanation for the missing mass.

Because of its consequence for the large-scale structure of the universe, the missing-mass problem also relates to other fundamental questions of cosmology. Even if we agree on what happened from the time the universe was a mere 10^{-32} seconds old to when the primordial hydrogen-helium gas was formed 300,000 years later, we still are faced with two questions. What happened *before* 10^{-32} seconds? And how did the primordial gas come to form stars and galaxies? It turns out that these questions are linked to one another, and to further questions. Why is the cosmic background radiation so very isotropic, varying by so little in its black-body temperature from one direction to another? Is the spatial geometry of the universe curved like the surface of a sphere, making it closed and finite? Or does it have the kind of curvature a saddle does, in which case it is open and infinite? Or might it be finely balanced between the two: flat, but open and infinite?³ Will it expand forever, or will the expansion eventually reverse, leading to a big crunch? These last questions require that we determine the value of the constant Ω we met above.

Although cosmological solutions to Einstein's equations do not require a cosmological term, they can easily be modified to accommodate one. Indeed, Lemître had allowed for one. He found a solution with flat space, the space of Euclidean geometry. And he also found two classes of solutions with curved spatial geometry, in one case open, in the other closed. It is difficult to determine from observations (based on counting the number of galaxies at increasing distances) which of these alternatives is true for our universe, but further data acquired in the last few years provide compelling evidence for flatness, or at any rate near flatness. The flat solution to the equations requires as input a very precise balance between the expansion initiated in the big bang and the gravitational attraction of the matter in the universe that acts to reverse the expansion. To achieve this balance in the absence of a cosmological term, an amount of matter is needed that is just such that the expansion of the universe, although continuing forever, slows down asymptotically to zero. Suitably expressed, this determines a *critical density* of mass (including the mass equivalent of energy—$E = mc^2$ yet again!). With a higher density than this, the universe is closed, eventually ceases to expand, begins instead to contract, and ultimately collapses in a "big crunch." With less than the critical density, the universe is open and expands forever, eventually expanding at a constant rate. A flat universe remains flat. But one that is not flat at the beginning never gets to be flat; it always diverges more and more from flatness. Since what we see now is so very close to a flat universe, it must have started out either *exactly* flat, or so incredibly close to it that in 13 billion years it is still as near to being flat as observations show. This alternative would require a very remarkable near cancellation, which, when we extrapolate back in time, looks extremely implausible.

It is not only matter that contributes to the overall mass-energy density. There is a contribution from the vacuum itself! One component of this comes from the cosmological constant, which may or may not be present. To this must be added a component that comes from quantum mechanics. In relativistic field theory the fields of all the elementary particles populate space. And even when the fields vanish on average, as in empty space, they are ceaselessly agitated by quantum fluctuations. These vacuum fluctuations may be thought of as giving rise to the creation and annihilation of particles, so-called "virtual particles" that never survive long enough to be detected directly. But virtual particles do contribute to the energy of empty space, with a positive contribution from some kinds of particle and a negative contribution from others. (Quantum fluctuations in the vacuum are not just theoretical phantoms; they shift the energy levels of electrons in atoms by tiny, but measurable, amounts. And on a more macroscopic scale, as shown by Hendrik Casimir, they generate a force between closely separated metal plates, and that too has been found by direct measurement to agree with his prediction.) On macroscopic scales, these contributions have effects similar to the cosmological term, and so may be included with it to give a revised value for Λ.

If one divides the contributions of matter and of the vacuum to the overall density by the critical density, one arrives at contributions to the omega parameter (called respectively Ω_m and Ω_Λ) that add up to a total that is less than 1 if the universe is closed, greater than 1 if it is open. And if the universe is flat, they have to add up precisely to 1. As noted above, it appears that the universe is very close to being flat. This means that the two terms add up to a total Ω very close to 1. But this conclusion leads to puzzles. The contributions to Ω_Λ from the quantum fluctuations of particle physics are hard to calculate, but the best estimates give numbers that are extravagantly large. It might be that there is a cosmological term that leads to cancellation. But that cancellation cannot be complete, since, as we shall see, recent observations imply a value for Ω_Λ different from zero. To obtain that value requires canceling all but 1 part in 10^{120} of the enormous number given by quantum fluctuations—an improbable degree of precise "fine-tuning."

But the story does not end there. Present observation suggests that even allowing for dark matter, the matter density is not that large, which means flatness requires a non-zero value of Λ. The values of Ω_m and Ω_Λ most favored by these observations are respectively close to 0.3 and 0.7, thus adding

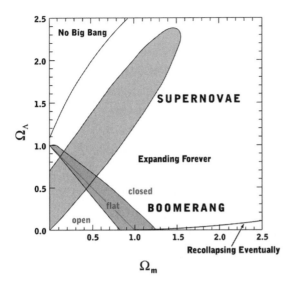

Fig. 11.3. Two complementary sets of observations help to determine fundamental cosmological parameters. Observations of supernovae and of the cosmological background radiation by Boomerang intersect in a region on this graph, indicating that the universe is flat and the cosmological constant is different from zero. (With thanks to Professor P.A.R. Ade)

up to a number close to 1, the value consistent with flat space (see figure 11.3). Although a flat universe stays flat, Ω_m and Ω_Λ have very different time dependence. As the universe expands, the matter density will fall, and the conspiracy that gives just the right sum for the two terms *now* will no longer work—unless their sum, Ω, has *exactly* the critical value 1. Anything else would suggest that there is something special about the present time, since observations now support Ω equaling the value 1. Since there's no reason to think that we *do* live at a special time—the Copernican principle in a different guise—theorists are tempted to believe in a model that has a flat universe now and forever and so has Ω equal to 1. In the early universe Ω_Λ is negligibly small, but approaches 1 as the universe ages and expands; conversely, at later times it is Ω_m which gets to be very small and Ω_Λ that approaches ever closer to 1. On a cosmological scale, the epoch in which Ω_m and Ω_Λ have similar values is very brief. And yet we seem to be living at just such a time. Is this just a coincidence? Or will cosmology provide an explanation?

The results for Ω_m and Ω_Λ, illustrated in figure 11.3, derive from two independent measurements. One of these (published in 1999) studied supernovae in remote galaxies. It is believed that all supernovae of a certain class have the same absolute luminosity, so that they can be used as "standard candles." The distance of each of the supernovae seen can be determined from the redshift of its parent galaxy. What was found was that these remote supernovae are fainter than expected, and that in turn implies that the rate of expansion of the universe is in fact speeding up. Ordinary matter or radiation always leads to slowing down of the expansion, but the Λ-term was introduced by Einstein precisely to counteract this—and the supernova data can be interpreted as evidence for a nonvanishing contribution of this kind. The gray elliptically shaped region in figure 11.3 outlines the most probable range of values for Ω_m and Ω_Λ consistent with these data. The supernova observations are sensitive to the difference between the two contributions to Ω, not to Ω itself, their sum, and rule out a zero value for the cosmological constant Λ. But they do not determine the value of Ω itself.

To pin down this remaining fundamental parameter that enters into the Friedmann-Lemaître-Robertson-Walker equation, we need data of a different kind. It turns out that different values of Ω lead to different predictions for the inhomogeneities in the density of matter at the time of decoupling from radiation when the universe was 300,000 years old. These become imprinted as tiny variations in the temperature of the cosmic background radiation across the sky. So a number of experiments have been planned to measure these fluctuations. The first of the infrared astronomical satellites to provide high-precision data on the cosmic background radiation was COBE (COsmic Background Explorer, launched in late 1989), which showed that the radiation obeyed to an extraordinary precision the Planckian distribution of intensity over wavelength. More refined observations by COBE were able to measure the departure from isotropy expected from the motion of the earth, and indeed the sun and the galaxy, through the universe, as well as minute fluctuations (at the level of 30 millionths of a degree!) in the temperature over and above that anisotropy. These observations confirmed that there were indeed density fluctuations at the time of decoupling but were not sufficiently precise to differentiate between conflicting cosmological theories, for example, to determine whether hot dark matter or cold accounts for the greater part of the missing mass. Nor could they determine the value of Ω.

In April 2000 the results of an experiment called BOOMERANG (Balloon Observations of Millimetric Extragalactic Radiation and Geomagnetics)

Fig. 11.4. MAP, the Microwave Anisotropy Probe. (Courtesy of the MAP science and engineering team; image at http://map.gsfc.nasa.gov)

were announced. This experiment is a microwave telescope that is carried to an altitude of 38 kilometers by a balloon. It was able to show to a high degree of accuracy that the universe *is* flat, so that Ω does have the critical value of 1. The results were confirmed a month later by a similar experiment called MAXIMA (Millimeter Anisotropy eXperiment IMaging Array). Further observations to be made by MAP, the Microwave Anisotropy Probe (figure 11.4), an orbiting satellite launched in the summer of 2001, will help to determine the details of the fluctuations of temperature in the microwave background still more precisely. This will lead to an even more accurate measurement of Ω, and taken together with the follow-ups to the supernova results of 1999, which showed that the expansion of the universe is accelerating, will give a still more precise determination of both Ω_m and Ω_Λ.

Fluctuations in density in the early universe are important for another reason. For such fluctuations are believed to have acted as the seeds around which stars and galaxies formed, condensing from the hydrogen-helium gas like raindrops in a cloud. We still do not know for certain whether stars formed first, later to aggregate into galaxies and then clusters, and eventually superclusters, of galaxies; or whether it went the other way round, with huge amorphous concentrations of matter clumping into protogalaxies from within which stars eventually formed. To be sure, we can observe the process of star formation in galaxies even now, but that still does not answer the cosmic chicken/egg conundrum.

It is clear, however, that the density fluctuations in the early universe are the clue. How did they originate? Why is the universe so nearly flat? Why is it that no matter in which direction we look, the cosmic background radiation has nearly the same temperature, even though it comes from sources so far apart that even in the 13 billion years of the life of the universe there has not been enough time for them to influence one another? Why are there so many more photons than protons? It turns out that all these questions turn attention to another. What happened before the universe was 10^{-32} seconds old? If we look back in time, retracing the steps we have followed, from today's universe, sparsely populated with clusters of galaxies of stars, back to the dense, hot gas that filled the universe when it was 300,000 years old; back further to the formation of protons and neutrons from the still hotter sea of quarks and leptons; then back yet further to the time when these particles were themselves created, we find that we must turn again to high-energy physics for guidance. At 10^{-32} seconds, the temperature corresponded to a typical interaction energy of 10^{13} GeV; this is an energy far beyond the reach of the most powerful particle accelerators or colliders. (CERN's LHC is designed to probe the physics of collisions at around 14,000 [i.e., 1.4×10^4] GeV). Even the highest-energy cosmic rays rarely exceed 10^{11} GeV. So our high-energy theory has to be based on an extrapolation from lower energies. It is therefore to some extent speculative, but nevertheless generally accepted.

The unification of the weak and electromagnetic interactions in the electroweak theory of Glashow, Salam, and Weinberg is associated with a symmetry that holds at high energies but is spontaneously broken at around a few thousand GeV. Their theory is one of the pillars on which the standard model of particle physics has been based, and its predictions have been tested with great precision. A more speculative unification brings together the strong interactions with the electroweak, in a symmetry that is manifest only at very much higher energies. This GUT (for Grand Unified Theory) is supported by less secure experimental evidence. One of the unanticipated consequences of the breaking of GUT symmetry is that it provides a mechanism for *inflation*. It is conjectured that, as the early universe cooled through the temperature at which the GUT symmetry breaks, the sudden change in the particle interactions this brought about was too fast for thermal equilibrium to be maintained. This resulted in a period of exponential growth, in which the visible universe expanded from around 10^{-31} meters in size to some ten centimeters across, all in 10^{-32} seconds. This expansion would have smoothed out any curvature in space, and so could account for the apparent flatness of the universe today. It would also explain the horizon problem—the puzzle of why cosmic background radiation coming from widely separated directions has very nearly the same temperature—since before inflation, these now-distant regions would have been close enough together for their conditions to be similar.

And there is another puzzle. Why is there more matter in the universe than antimatter? Or put another way, why is it that protons and electrons are common but antiprotons and positrons rare? Andrei Sakharov was the first to recognize, in 1967, that three conditions are needed for the emergence of this imbalance from a prior state of symmetry.[4] One was the violation of the C and CP symmetries of particle physics (see chapter 8), already known from high-energy physics experiments. Another was the violation of one of the conservation rules of the standard model, so as to allow protons to decay, for example, into positrons. This has never been seen to happen, and it certainly proceeds at a *very* slow rate, if at all. After all, we know that the protons in the stars have lived as long as the universe, and a more precise experimental limit on the average lifetime of a proton sets it to be greater than 10^{32} years, many orders of magnitude greater than that![5] Nevertheless, proton decay is not ruled out, and is in fact predicted by the very same GUT theories as are used to explain inflation.

Sakharov's third condition for the prevalence of matter over antimatter was an episode in the evolution of the universe in which it departed from thermal equilibrium—and that is just what happens in inflation. This same departure from particle/antiparticle symmetry can also account for the ratio of photons to protons that we observe today. It is worth noting that the exponential expansion of the universe during inflation is just the sort of behavior described by de Sitter in 1917!

There's more to inflation than I have been able to cover in this short account, and many of the details of the idea need to be tested and refined both for their theoretical consistency and their observable consequences. Suffice it to say that most cosmologists accept *some* sort of inflation episode as a necessary component of the big bang model, if that model is to account for what we observe in the universe today.

There remains one further question to be addressed, which I have raised but not yet answered. What happened at still earlier times, before inflation? When the universe was only 10^{-43} seconds old, when its size was a mere 10^{-35} meters across, the very notions of space and time become blurred. This is because at these scales it is no longer possible to ignore the effects of quantum mechanics. Einstein's equations no longer serve to describe spacetime. We need a quantum theory of gravity, maybe even a quantum theory of cosmology. The most promising candidate for a consistent quantum theory of gravity has come from superstring theory and is one of its outstanding successes. But its implications for the event we call the big bang are still controversial and unresolved. It is, however, generally accepted that quantum fluctuations in the spacetime geometry would generate corresponding fluctuations in density at the beginning of inflation. And these would be "frozen" or "imprinted" during inflation and thus persist, but now enormously enlarged in scale, even after inflation ended. The galaxies and galaxy clusters were seeded by these fluctuations. They have left their mark as the skeins and voids in the distribution of galaxies and in the tiny variations from one direction to another in the temperature of the cosmic background radiation.

Even when quantum effects are added to the classical picture, the big bang cosmology still leaves some deep questions unanswered. The most problematic of these is the physics of the big bang itself. The classical picture of expansion from a primordial big bang can be illustrated by an even simpler model than the expanding balloon of figure 11.1: let us represent the universe by the circumference of a circle that expands with time as the universe ages. Then the sequence of these expanding circles can be arranged to generate a cone, where the axis represents time (see figure 11.5). You may find it

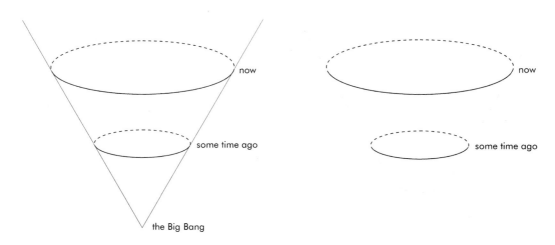

Fig. 11.5. Another model of the expanding universe. The surface of the cone represents spacetime. The circular cross-sections represent space at different epochs. At the vertex of the cone—the big bang—space is shrunk to a point. This is a singularity in spacetime. In Hawking and Hartle's theory, this singular point is rounded off, as suggested in the second version.

difficult to do so, but you must ignore the fact that your cone appears to be "immersed" in the familiar three-dimensional space of everyday: one of these dimensions is being used to represent time. The apex of the cone, which represents the big bang itself, is a singular point, a point that marks the beginning of time, at which the density of matter in the universe was infinite. One might conclude that the laws of physics break down there, leaving us with the unanswered question, How did it all begin? One intriguing speculative theory suggests that there was no singularity, and that the laws of physics can still be applied to the big bang itself. This is the "no boundary" proposal of Stephen Hawking and James Hartle. It builds on a concept already widely used in quantum field theory and string theory, the notion of *imaginary time*. When making calculations in field theory, it is often useful to pretend that the time variable is pure imaginary, as was done by Minkowski when he introduced the idea of spacetime. This artifice makes many steps in the calculation easier to formulate, and also mathematically more secure. Only at the very end of the calculation is the time variable made real once more. Hawking and Hartle have suggested that we regard this as more than just a mathematical trick. They argue that when portrayed in imaginary time, spacetime has no boundary: like the surface of a sphere, it is finite but without boundary. A point, but not a singular point, represents the beginning of the universe. As Hawking puts it, it is an ordinary point of space and time, as the North Pole is an ordinary point on the earth. To be sure, the meridian lines all meet there, but that's just a consequence of the way we draw our maps. Hawking and Hartle then suggest that the laws of physics can be applied just as well at this point as at any other, and "the universe would have expanded in a smooth way from a single point." To be sure, if we revert to using *real* time, there was a singularity, at which the laws of physics would have broken down. But the way the universe began can still be described by the laws of physics, since we can do the calculations first with imaginary time, where there is no singularity, and at the end of the calculation replace imaginary time with real time. According to this argument, the question "What came before the big bang?" is based on a misunderstanding. As Hawking says, it is rather like asking what is north of the North Pole.

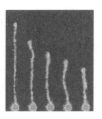

12

DOWN TO EARTH

★ ★ ★

Physics on a More Human Scale

LET US SUPPOSE THAT M-THEORY REALLY DOES FULFILL ITS PROMISE, and that some time soon we will have a unique, "rigid" theoretical basis for our understanding of the fundamental laws of nature. Our present standard model will then be seen as an emergent theoretical framework, appropriate for the description and understanding of high-energy particle physics. Its parameters will no longer be arbitrary, "fed in by hand" from outside the theory and adjusted to fit the data, but will be determined uniquely from the underlying fundamental theory. Of course, this determination may only be possible in principle, because the calculations may prove to be impracticable or unrealistically difficult. Nevertheless, suppose we knew that they *could* be done, with unambiguous results. Would physics then come to an end?

Of course not! Chemistry did not become moribund just because its underlying physics could be completely formulated through the laws of quantum mechanics.[1] Most physicists work on problems I have not even mentioned in this book, and many of the most significant advances of the twentieth century have been made in areas of research far removed from high-energy particle physics, and even further from superstrings or M-theory. It may be comforting to know that "in principle" quantum electrodynamics provides the basic laws from which an understanding of solid-state physics can emerge. But it does not help us to give an adequate account, for example, of superconductivity. A superconductor has no electrical resistance. Kamerlingh Onnes discovered the phenomenon in 1911 when he cooled mercury to below 4 K, but it was not until 1956 that John Bardeen, Leon Cooper, and John Schrieffer offered an explanation of its physical mechanism: the BCS theory. In some cuprate ceramics, superconductivity occurs at the much higher temperature (77 K) at which nitrogen liquefies. It is believed that the BCS mechanism does not apply to these HT_c (high critical temperature) superconductors, and there is still no general consensus on the correct alternative explanation. If we *did* have a better understanding of what causes their superconductivity, we might be able to devise materials that were superconducting at room temperature, and that would have an immensely important impact on electrical technology. But this better understanding will certainly not come from quantum electrodynamics!

The electron was discovered only a little more than a century ago. Faraday's experiments and Maxwell's theory of electromagnetism were arguably of more significance for human society than any nineteenth-century political revolution.[2] Electric power and light at the touch of a switch have transformed our lives. The versatile electron is not only the agent of these democratizing forces but also of electronics, which has driven forward the revolution in communications, bringing the remotest regions of the world instantly into our living rooms or offices. Like Aladdin, we have the genie of the lamp at our command. Do not blame the physicist for the stupidity of some of the things we ask for! The electronic revolution depends on the quantum properties of solids, on the special features of the energy levels in crystalline silicon, for example. Our ability to store information on the discs of computers in ever more compressed form is based on the improved understanding of magnetic materials, and magnetism is itself an intrinsically quantum phenomenon. An inexpensive computer of the kind found in many homes today has more computing power and a bigger memory than those which guided the first landings of men on the moon. Our homes receive entertainment and information from international communications networks, linked by glass fibers carrying a multitude of messages, and with an efficiency and economy, inconceivable without this enabling technology. As a matter of fact, the World Wide Web was developed at CERN to facilitate the transfer of data between particle physics research groups working together on large collaborations. It now gives millions access to cyberspace shopping malls, more information than they would find in their local public library—and, should they want it, data from physics experiments.[3]

I want to give you some impression of the wide range of exciting research in physics today in areas that are not at the frontiers of subnuclear physics nor of astrophysics or cosmology. There are other frontiers besides these, either of extreme conditions or of extreme precision. This is the physics done in university laboratories rather than at huge international facilities housing accelerators or telescopes. Here too one can witness the interplay between technological advance and scientific discovery, each feeding on the other. Otto von Guericke's seventeenth-century pumps have long been superseded; James Dewar required better vacuum technology for his invention of the vacuum flask, the familiar thermos bottle of the picnic basket. Dewar's flasks are also the basis for cryogenics, the study and application of low temperatures which led Onnes to the liquefaction of helium and thence to his discovery of superconductivity. Today we need, and can achieve, pressures as low as a millionth of a millionth of atmospheric pressure for some branches of physics.

As in the past, new technology and novel applications are the basis of most experimental discoveries, which in turn provoke and authenticate theoretical advances. Sometimes there is a very rapid transfer of a frontier technology from the research laboratory to industry and everyday use. But it can often take decades for applications to be realized. The theory of stimulated emission, or which laser technology depends, was initiated by Einstein in 1917 in one of his most profound contributions to quantum physics; ironically, he was never reconciled to the break with classical determinacy which his work helped to promote. It was not until 1960 that the first optical laser was demonstrated.[4] For some years the laser was often described as a solution looking for a problem, but now lasers have an essential place in research laboratories, as well as in countless other applications, whether in surgery, in barcode readers at supermarket checkouts, or in scanning the records in CD players. This time lag from fundamental science to multibillion-dollar industry is not unusual, and it should sound a cautionary note to those who expect rapid return from research investment. Einstein's research was not in any sense "mission-oriented," but it paid off in the end!

Lasers have had a very considerable impact on fundamental research, as well as on shopping and home entertainment! The light from a laser is highly monochromatic, that is to say, it has a sharply defined wavelength; and it is coherent, the waves remaining in step with one another both in space and in time. Laser light can also be extremely intense, and laser pulses can be extremely short in duration. It is these features, above all, that make lasers such invaluable tools for physics research and that allowed the development of a whole new branch of physics, quantum optics. Spectroscopy has a long history of application in atomic and molecular physics. The interaction of an atom with light depends on the internal structure of the atom or molecule: it can only absorb a photon having just the right wavelength to produce a quantum jump to a higher energy state. Likewise, if it subsequently emits a photon, the wavelength will depend on the quantized energy levels of the atom or molecule, and these energy levels are related to its structure. The precision of lasers has opened up new applications: for example, in environmental studies, where they are used to detect trace contaminants at very low concentrations. Since the wavelength "seen" by a moving atom will be Doppler-shifted, laser spectroscopy can differentiate between atoms in motion and at rest and can be used to study the kinetics of chemical reactions.

Lasers are also used to trap single atoms or very tenuous clouds of atoms and bring them almost to rest, so enabling them to be studied with much higher precision than when they are close to one another, as in a solid or liquid, or even a gas under normal conditions. At room temperature, the atoms and molecules in the air are moving in different directions with speeds of around 4,000 kilometers an hour, colliding with one another as they do so. At lower temperatures, their speed falls, but at low temperatures gases condense to liquids or solids, and the atoms interact more strongly with one another. By cooling a gas at greatly reduced pressure, it is possible to inhibit condensation to the point where one can begin to study atoms without the interferences of frequent collisions and interactions. But they are still moving in all directions, and in consequence the spectral lines are broadened through the Doppler effect. One would like to slow them down as much as possible, and laser methods developed by Steven Chu, Claude Cohen-Tannoudji, and William D. Phillips have achieved that objective, cooling gases to within a few millionths of a degree above absolute zero.

If a laser is tuned to a wavelength just above that at which it would be absorbed by an atom at rest, a photon in the laser beam can only be absorbed by an atom moving toward the laser, because it is then "seen" by the atom as blue-shifted to a lower wavelength by the Doppler effect. The atom is then slowed down, and even though it will recoil and so pick up speed when it emits a photon, the recoil direction is random. The net effect is to reduce the average speed of atoms moving toward the laser. Chu and his collaborators stopped a beam of sodium atoms entering a vacuum chamber by directing the atoms head-on toward a laser beam, and then conducted them to a trap. This was made by arranging three pairs of lasers at right angles to one another, with their beams all directed to cross at the same point. The result was that the average speed of all the atoms in the region where the beams crossed was further reduced, and the cloud of sodium atoms was thus captured. Any atom moving so as to escape from the trap encountered a laser beam that pushed it back. And where the beams intersected, the atoms slowed down as though they were moving in treacle: hence the name *optical molasses* for such an arrangement.

The trap so formed is leaky, because the atoms fall out of it—quite literally, through the force of gravity. To retain them longer than about a second required another invention, which introduced a magnetic field into the laser system, and this could be arranged to provide a force that balanced the weight of the atoms. Further refinements by Cohen-Tannoudji and by Phillips have allowed temperatures as low as 0.2 microkelvins, two-tenths of a millionth of a degree above absolute zero. At this temperature the (helium) atoms used have a speed of only 2 *centimeters* a second. (Compare this with the average speed of atoms in a gas at room temperature, which is 4,000 kilometers per hour—figure 12.1.)

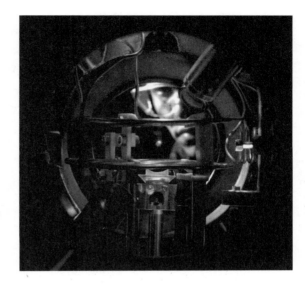

Fig. 12.1. This photograph shows about a million atoms of sodium held in a magneto-optical trap of the kind described in the text. (Courtesy of the National Institute of Standards and Technology, Gaithersburg, Maryland)

The exceptional coherence of laser radiation is exploited in interferometry, extending the range of applications of what was already a powerful technique. In an interferometer, such as that used by Michelson and Morley in their attempt to detect the motion of the earth through the ether, a beam of light is split in two (for example, by a half-silvered mirror), and the two components are combined together again after having followed different paths. The extent to which the two components now reinforce one another or cancel out provides a sensitive measure of the path difference, to within a fraction of the wavelength of the light. The path difference over which the interference effects are visible is limited by the distance over which the light remains coherent. Hence the benefit gained from the use of laser light, since coherence now extends over as much as 100 meters, rather than the few centimeters which is the best that can be obtained otherwise.

In Canterbury, New Zealand, a group led by Geoff Stedman has developed a ring laser, which reflects beams around a square, one going clockwise, the other anticlockwise. The resultant interference pattern—11 million nodes and antinodes spaced around the ring—remains stationary with respect to the local inertial frame of reference. But the earth's rotation carries the body of the ring around in that frame, so the nodes and antinodes pass by the mirrors and a detector, generating a beat frequency in the light emerging from the ring. The result was to give a frequency of 69 Hertz (79.4 Hz for their more refined instrument, which came into operation in 1998). When this is fed into a loudspeaker, you can "hear" the rotation of the earth. The Canterbury system can be thought of as a ring laser gyroscope, and it is far more sensitive than the gyroscopes used for navigation in aircraft, better in fact than any other technique yet employed to sense the rotation of the earth, and able to rapidly detect the minute changes in its rate caused by seismic and tidal events.[5]

Another advantage of lasers is the sharp directionality of their beams; they hardly diverge at all. Even after traversing the 5 million kilometers separating the spacecraft that will carry the LISA interferometer (see chapter 11), the beam from one to another can still be detected. It is designed to detect changes in the distance separating the spacecraft to an accuracy of 2×10^{-11} meters (that's better than 1 part in 10^{20}!), the degree of precision needed to detect gravitational waves such as might originate from the vicinity of a black hole at the center of the Galaxy.

This is surveying on a grand scale. Lasers have more mundane applications as tools for surveying, but they are impressive, nevertheless. The distance to the moon was measured to within about 15 centimeters by measuring the time it took for a laser pulse to be reflected back to Earth from corner reflectors placed on the moon by Apollo astronauts. Other applications which depend on the high intensity and directionality of laser light include *optical tweezers* and *optical scissors*. A tightly focused laser beam will attract to its focus small transparent objects, and they can then be moved about as though held by tweezers. Single molecules can be "labeled" with a fluorescent marker and attached to tiny particles, less than a thousandth of a millimeter across. These can then be trapped in the laser beam of optical tweezers and moved about by sensitive and accurate piezoelectric controls.[6] In this way we now have the capacity to manipulate individual molecules and can, for example, measure the force needed to separate the strands of the double helix of DNA, or the force exerted by a single molecule of myosin as it pulls itself along an actin filament in the process of muscular contraction (plate 18).

The intensity of laser light can be used to slice through sheets of steel—as everyone who has seen a James Bond movie will know. But it can also perform far more delicate operations. By focusing laser light down onto an organelle (e.g., a mitochondrion or a chloroplast) in a single cell, one can conduct exquisite microsurgery, without harming the rest of the cell. Optical scissors may be used in this way to cut apart chromosomes inside a living cell.

The laser in a CD player is a solid state laser, different from the gas lasers often needed for research. Solid state lasers are also used to launch the light that supports the digital signals used in optical telecommunications, light transmitted down glass fibers which now carry a large fraction of all the telephone traffic of the world. Optical links have many advantages over old-fashioned copper wires: the pulses do not spread into one another as they travel through the fibers; they can be amplified so as to maintain the signal over large distances; and many messages can be compressed into the same channel. Opto-electronics is still relatively in its infancy for computing and data processing applications, but I expect that soon it will begin to challenge silicon-based technology.

Another kind of interferometry, using atoms rather than photons, has only very recently become possible. Just as photons are the particlelike manifestation of something that was classically viewed as a wave, there is a complementarity between the particle- and wavelike properties of things like electrons. But also of things like atoms, or even molecules. Atom interferometry exploits the wavelike properties in a beam of atoms. Until recently, we did not know how to make optical elements—mirrors, for example—that did not destroy the coherence of the quantum waves needed for interference, but that technological barrier has now been overcome. Since the length of the quantum waves of atoms is so much smaller than that of light, atom interferometers have the potential to improve on the sensitivity of optical ones by as much of a factor as 10^{10}. Although still in the very early stages of development, an atom-interferometer gyroscope has been built by Mark Kasevich and his collaborators at Yale, which can detect rotations a thousand times slower than that of the earth on its axis.

The ability to control and manipulate single atoms has had a profound impact on fundamental physics and in particular has opened up new ways to address old questions. For example, the "shifty split," mentioned in chapter 4, at the heart of the interpretation of quantum mechanics has been probed as

Fig. 12.2. Sir Charles Babbage, the progenitor of the programmable computer. (Portrait by Laurence, reproduced with permission from the Master and Fellows of Trinity College, Cambridge)

never before in the experiment by Serge Haroche at the Laboratoire Kastler-Brossel in Paris. Instead of Schrödinger's thought experiment, we can now do real experiments: an atom in a microwave cavity—closely analogous, to his cat in a box—can be used to measure and follow the decoherence induced by interactions with the environment, and so test the widely held view that it is these which bring about the transition from quantum superposition to classical certainty.

Decoherence brought about by interactions with the environment are, on the other hand, to be avoided at all costs if one wishes to exploit another wonderful advance made possible by quantum mechanics. This is the quantum computer. Ever since Charles Babbage's (figure 12.2) analytic engine and its descendant, the modern computer, the approach to computation has been essentially linear, step by step. To be sure, we have parallel processing machines, but these are, in essence, devices that allow a large number of linear machines to run side by side. In a quantum computer, many computations would be conducted simultaneously with the same basic arithmetic unit. By making the initial data on which the calculation is to be performed not just one string of numbers but a quantum superposition of many numbers, the same operation can be performed on all of them simultaneously. The theory is already well advanced, and the first halting practical steps have been taken. This is another area where I would predict that within a decade, or more likely two, the technical obstacles will have been overcome, with very important consequences. The speedup to be expected from a successful implementation of quantum computing is not just an incremental step along the path we have been following for the past five decades; it opens the prospect of performing challenging but simple arithmetical tasks in a very short time. The security of many coded transactions—for example, financial transfers between banks—depends on cryptography, which in turn depends on the extremely difficult task of finding the prime numbers factorizing a very large product. If this could be done at the speeds theoretically suggested for a quantum computer, the code could be broken in hours rather than on the time scale presently relied upon for the security of the key—which is millions of years.

Fortunately, our credit cards will be safe: quantum mechanics can come to the aid by offering a more secure form of cryptography. Through the weird consequences of quantum superposition, not only can information be transmitted securely, but the legitimate users of the system would know at once if some outsider had tried to tamper with it. For the very act of tampering would destroy the coherence on which the method depends.

Quantum coherence is the basis of another important recent advance. As the temperature of a system falls, so does the average kinetic energy of the random motion of its atoms or molecules; indeed, temperature is just a measure of the energy of this random motion. At low temperatures quantum effects begin to manifest, because particles of the same kind are indistinguishable one from another. All electrons, for example, are alike. But electrons are fermions, which means that they obey the Pauli exclusion principle: no two of them can simultaneously occupy the same quantum-mechanical state. So at the very lowest temperature, at absolute zero, electrons will "fill up" the available states, starting from those of lowest energy and proceeding up the energy ladder till all the electrons are accommodated. Bosons behave differently, since the Pauli exclusion principle does not apply to them: as the temperature drops, more and more of them occupy the state with the lowest energy level. When there are a macroscopic number of bosons in this one quantum state, all in effect having the same wave-function, the system behaves in a remarkable fashion. For example, in liquid helium below 2 K, a substantial fraction of the atoms (which behave as bosons) are all in the same state. The liquid flows without friction; it has become a *superfluid*. The theory of superfluids owes much to the work of Lev Landau, who was awarded the 1962 Nobel Prize for his discoveries. The BCS theory of superconductivity is related to this work, because it proposes that electrons in a superconductor can be bound together in pairs, which then behave as bosons and in effect become a superfluid, carrying the persistent current that characterizes superconductivity.

But in superfluids and superconductors, the bosons still interact with one another. The challenge was to get bosons to behave in this way in a situation where they were so far apart from one another that their interactions were negligible. They would then become a "Bose-Einstein condensate," first described in 1925 by Einstein. This has been called the fifth state of matter (the others being the familiar gas, liquid, and solid, and the less familiar plasma). At last in 1995 the conditions for making a Bose-Einstein condensate were achieved, when around 2,000 rubidium atoms were cooled in a trap to within less than 170 billionths of a degree above absolute zero. As Eric Cornell and Carl Weiman, who performed the experiment, said, they had produced the lowest temperatures anywhere in the universe—unless some physicists in another solar system had done the same kind of experiment! The study of this exotic form of matter is a burgeoning area of research. For example, light can be slowed down enormously by shining it through a Bose-Einstein condensate. A team led by Lene Hau at Harvard University and the Rowland Institute for Science has slowed the speed of light to 38 miles per hour! Hau says that she believes that communications technology, television displays, and night-vision devices will soon be improved because of her team's research.

In 1959 Richard Feynman, one of the pioneers of quantum electrodynamics, gave a talk at a meeting of the American Physical Society at the California Institute of Technology, a talk now regarded as a classic. It was called "There's Plenty of Room at the Bottom: An Invitation to Enter a New Field of Physics." "This field is not quite the same as the others," he pointed out, "in that it will not tell us much of fundamental physics [. . .] a point that is most important is that it would have an enormous number of technical applications." He was referring to what is now often described as *nanotechnology*, the physics and engineering of devices on the scale of just a few atoms. "In the year 2000, when they look back at this age, they will wonder why it was not until the year 1960 that anybody began seriously to move in this direction." He offered two prizes of $1,000 each, with the expectation that he would not have to wait long for claimants. One was for the first person to make an operating electric motor that could fit in a 1/64-inch cube. That prize was won in little over a year. The other was to reduce the information on the page of a book to an area 1/25,000 smaller in linear scale in such a manner that it can be read by an electron microscope. That prize has not yet been claimed—but it will be soon!

The successful drive to make ever smaller devices for electronics is responsible for the spectacular advances in computer design and manufacture, as more and more components are crammed onto silicon chips. The microprocessor pioneer Gordon E. Moore, cofounder of Intel Corporation, predicted in 1965 that the density of transistors on the semiconductor chips used in computers would double roughly every two years. "Moore's Law" has held pretty well up to now, but it can't go on indefinitely, since the size of atoms sets a lower limit to the smallest conceivable transistor, and that limit will be reached in the first decades of this century. The computer manufacturing industry has its sights set on new technologies that will move from the planar geometry of today's chips into the third dimension, assembling circuits and devices in multilayered structures that will compress yet more computing power into yet smaller volumes. To do this, they are turning to nanotechnology. But the design of ever smaller systems will require more than just a change of scale. As the size of the system is reduced, quantum effects become ever more relevant, and the physical behavior of the system becomes qualitatively different. A wide range of studies on MEMS (MicroElectroMechanical Systems) may be needed to implement these novel kinds of computer manufacture and to solve other problems encountered in the drive toward miniaturization.

A related area of rapid advance is the development and application of *scanning probe microscopes*, which can examine structures as small as an atom. Optical microscopes are limited in their resolving power by the wavelength of light, around 500 nanometers (5×10^{-7} meters). But atoms are a thousand times smaller than that. Electron microscopes can explore structures as small as a few atoms across, but it is difficult to look at materials interesting to biologists at this resolution, because the electron microscope requires the sample to be dry and in a high vacuum, and preparing the sample can destroy or distort what was to be examined. The new microscopes operate on an entirely different principle, by scanning across the surface of the sample to be imaged with an extremely sharp, pointed probe. One of a number of techniques is then used to "sense" the surface as it is scanned in a raster pattern like that on a TV screen, and the result is then processed by computer to provide an image. The atomic force microscope, for example, senses the force exerted as the probe is pulled across the surface to be imaged, thus detecting the topography of its minute hills and valleys. In a scanning tunneling microscope, on the other hand, the probe is kept a short distance away from the surface, and an electric potential is maintained between the probe and the surface (which has to be electrically conducting). Even though there is no electrical contact, a current can flow across the gap—a consequence of the quantum-mechanical "tunneling" of electrons across what in classical physics would be an impassable barrier. This tunneling current is then detected and used to sense the surface. Variants on this very versatile technique are already commercially available and are still being refined.

The new microscopy is but one of an emerging range of techniques that are already making the transition from research laboratory to industrial exploitation and that rely on the ability to manipulate and control individual atoms, to probe materials on an atomic scale, and to fabricate devices of extraordinary refinement and precision. The properties of tiny electronic devices, so small as to involve only a few electrons at a time, can be radically different from the transistors and switches on a grosser scale. With the increased understanding of phenomena on the atomic scale and the ability to handle individual atoms and molecules has come the promise of emulating some of the activities of the molecules in living systems or of interacting directly with biophysical systems. We can now make chemical sensors (artificial noses!) that respond to specific individual molecules; we may soon be able to use bacteria to "grow" electronic devices. When working at these dimensions, from a billionth of a meter (a nanometer) and even smaller, one's intuition has to be modified: quantum effects begin to dominate, and the behavior of matter is not just a scaled-down version of its behavior in larger aggregates.

We can learn a great deal from the study of individual atoms, and in many ways this line of research is attractive, because it deals with simpler things than the messy aggregates of atoms and molecules we encounter on the everyday scale. But many of the phenomena that occur on the larger scale are not evident at the smaller. A whole new science has emerged that is concerned with the behavior of *complex* systems, whether they be large aggregates of simpler atoms or communities of individual animals or human societies. It is not clear to me that this study of complexity will yield universal principles applicable to such a broad diversity of systems, but I have no doubt that we are challenged to understand how complex behavior emerges from the interaction between large numbers of simple components, and that progress in meeting that challenge will be a feature of twenty-first-century science.

This area of research is essentially multidisciplinary. It is remarkable that the behavior of systems as diverse as those in ecology, anthropology, chemistry, economics, and political science should have features in common. What many of them seem to share is scale-invariant (fractal) geometry and ideas of "self-organized criticality." These ideas have applications also in physics—for example, to the shape of sand dunes, diffusion, and phase transitions—which is why physicists are able to contribute so substantially to this research.

Many physics problems of enormous difficulty impinge directly on human concerns. Our way of life in the industrialized countries is heavily dependent on an abundant and inexpensive supply of energy. We are profligate in its use, but even if we learned to husband fossil fuels more carefully, even if we could calm the fears concerning nuclear energy, and especially the safe disposal of radioactive waste, there would still remain a gap between the burgeoning demands of developing countries and the supply of fuel they need to catch up with their expectations. Much hope has been placed in the future

Fig. 12.3. Inside JET's toroidal vacuum vessel. The height of the inside of the vessel is 4.2 meters. (Courtesy of the European Fusion Development Agreement-JET)

development of controlled thermonuclear fusion to supply pollution-free energy from hydrogen, abundantly available to all. I am skeptical, because this seems to be an ever retreating horizon. As the old joke says, "Fusion is the energy source of the future . . . and always will be." The fusion of hydrogen to make helium is what generates the energy radiated from the sun. To make it happen requires conditions similar to those in the sun, a combination of high temperature and high pressure. The most promising approach to achieving the critical threshold for fusion reactions remains that pursued in generators that use magnetic fields to confine, compress, and heat a plasma. Progress has been made by the Princeton Plasma Physics Laboratory, both with their TFTR (Tokamak Fusion Test Reactor), which achieved 10 megawatts of power and a temperature of 500 million degrees before it ceased operation, and from the subsequent National Spherical Torus Experiment. Another device, operated by an international collaboration of European countries is the JET (Joint European Torus, operated by the U.K. Atomic Energy Authority), which has generated a peak power of 16 megawatts—but even at a more modest 10 megawatts, the power output was sustained for only about half a second, and more power was needed to operate the machine than could be extracted from it (figure 12.3). Nevertheless, this does represent a substantial advance, and I would like to be proved wrong in my skepticism concerning fusion as a source of energy. Notwithstanding the United States' withdrawal from the project, there is still support for ITER, the proposed International Thermonuclear Experimental Reactor collaboration, which will probably now be based in Canada.

Strictly speaking, energy is free; it is only free energy which is expensive! Energy is never lost, just converted from one form to another. The energy released in thermonuclear reactions is just the energy content of the mass lost in changing hydrogen to helium. Only *free energy* is useful, available to do work. The difference between energy and free energy involves entropy. We really ought to be more entropy-conscious! That's really what improving the efficiency of machines is all about, and physics research can guide practice and assist progress. Similar problems confront us in global warming, atmospheric pollution, and a whole raft of issues that involve thermodynamics and that impinge in a very direct way on society. *How* we choose to address those problems is primarily a political question, but it will be physics that sets the range of options before us and alerts us to their possible consequences.

13

EPILOGUE

★ ★ ★

Into the Unknown

AS NIELS BOHR SAID, "PREDICTION IS HARD: ESPECIALLY OF THE future." But I have, perhaps rashly, made some predictions about what the twenty-first century will bring. The agencies that fund research demand "road maps" and "milestones" to chart out the new territory into which physics explorers wish to venture. Speculation is part of planning, and the costly experiments of particle physics or space-age astronomy have to be planned decades in advance. This means that in these areas the principal research objectives for many years ahead have already been scrutinized and screened by committees and funding agencies.

So it is with some confidence that I can assert that the Higgs boson will be discovered at the Tevatron collider at Fermilab (the United States national accelerator laboratory in Batavia, Illinois)—or if not, that some alternative explanation for the successes of the standard model will emerge.[1] In either case, I expect that by 2020 we will have a pretty good understanding of the origin of mass, a question closely related to the role of the Higgs boson in the standard model. The BaBar experiment, as well as a successor planned for the LHC, will provide the crucial data needed to understand the details of CP violation and hence the origin of the asymmetry between matter and antimatter in the universe. One of the predictions from QCD, the theory of strong interactions, is that a quark-gluon plasma can be formed at high temperatures, a state of matter that is relevant to the understanding of neutron stars and of what happens in a supernova explosion. It is also believed that in the very early universe a quark-gluon plasma was the state of the strongly interacting particles, and there may even be stars in which this exotic form of matter dominates. Experiments have already given tantalizing glimpses of its formation in high-energy collisions between heavy ions,[2] and in the next few years I expect that much more will be revealed.

Theory suggests that the fundamental particles and their interactions exhibit supersymmetry, although it is a broken symmetry and therefore hidden from our present experiments. I believe that by 2010 there will be firm evidence to support the theoretical predictions, and that the first of the supersymmetric partners to the quarks, leptons, and bosons of the standard model will have been identified. The lightest of these, the neutralinos, are the favored candidates for the WIMPs that cosmologists believe are the particles of cold dark matter needed to provide the missing mass in the universe. I predict that they will be discovered deep in a mine in Yorkshire, England, where a team of astrophysicists and particle physicists are searching for them.

The masses of all three kinds of neutrinos will be known by 2010. Data from the MINOS experiment, which will explore neutrino oscillations by directing a neutrino beam from the Fermilab accelerator in Illinois toward a detector in Minnesota,[3] will play a vital part in this determination of neutrino mass. Massive neutrinos now sit uncomfortably within the standard model, but by 2010 we will have adjusted it so as to retain its best features while accommodating the new data. It has been suggested that this may best be done by uniting the symmetries in a Grand Unified Theory. I will stick my neck out, and guess that M-theory will lead us further along the way beyond the standard model, and it will explain why neutrinos have mass, how they can be accommodated in a GUT—and also why there are just three generations of quarks and leptons.

The World Wide Web was first developed at CERN as a tool for collaboration in the high-energy physics community. But when the LHC comes on stream in 2005, vastly improved facilities will be needed. The data generated by the LHC will be about 1 petabyte (10^{15} bytes) of data per second—equivalent to filling a stack of CD-ROMS one mile high every second—from each of its four detectors. This enormous amount of data has to be available for analysis by thousands of scientists spread across the globe. Current computing technology does not scale to handle such extremely large data flow rates nor the complexity of the analysis process. A new global network, called GRID, is being developed that will not only provide for the data distribution and processing technology demanded by high-energy physics but will also underpin research in other areas, such as genomics and earth observation. And no doubt it will provide the technology and software for the next generation of the Internet.

In Antarctica a telescope sunk a kilometer deep in the ice to detect neutrinos will open a new window on the universe. It is made from hundreds of photomultiplier tubes, which will catch flashes of light in a cubic kilometer of ice to identify neutrinos that have come right through the earth and to establish the direction of their source. This is the Ice Cube telescope. (A much smaller version, known as AMANDA (Antarctic Muon and Neutrino Detector Array), came into operation early in 2000.) Just as x-ray and gamma-ray astronomy have complemented optical and radio astronomy, neutrino astronomy will bring new information from the deepest regions of space.

The microkelvin fluctuations across the sky in the temperature of the cosmic background radiation will be measured with improved precision and will explain the origin of galaxies from quantum fluctuations in the immediate aftermath of the big bang—or an alternative theory will be put forward, generating a need for other observations. Gravity waves will be detected at LISA or LIGO—or if not, one of the central predictions of general relativity will fail, and that is something I believe to be unlikely in the extreme. I predict that we will have learned the origin of gamma-ray bursters by 2005. Dark matter searches will have revealed the nature of the missing mass, the clue to resolving the puzzles posed by the motion of stars within galaxies and the cosmological questions on the birth of galaxies and the ultimate fate of the universe. Besides the neutralinos being sought in the Yorkshire salt mine, there might be *axions*, much lighter particles predicted by theorists to explain the lack of CP violation in the strong interactions. Theorists predicted that axions could be formed when the primeval quark-gluon plasma underwent the transition to the state of nuclear matter we observe today, in which quarks and gluons are bound inside strongly interacting particles. Experiments to look for axions are in progress at the Lawrence Livermore National Laboratory in the United States and in Kyoto in Japan. Neutralinos and axions are candidates for *cold* dark matter, but it is still believed that a small amount of *hot* dark matter is needed to account for the formation of superclusters of galaxies from density perturbations in the early universe—and that is a possible role for massive neutrinos to fill.

But even when the missing mass has been accounted for, when the dark matter has been identified, there is still the mystery of the accelerating expansion of the universe. Is there a cosmological constant? Or is there a new kind of field that generates a universal repulsion, just as gravity generates universal attraction? Some cosmologists suggest that the universe is filled with what they have called "*quintessence*."[4] This would behave rather like the lambda term in the cosmological equations, but would not be a constant and could change over time. More data on the density fluctuations in the early universe will come from satellite missions such as MAP, and together with a more convincing theory, these will decide between a cosmological constant, quintessence, or some other explanation for the acceleration of the Hubble expansion. I expect that we will have the answer by 2010.

The beautiful pictures of galaxies at the furthermost limits of observation that have been captured by the Hubble telescope will be supplemented by others from new ground-based instruments, using interferometry to give angular resolution 100 times better than the already impressive capabilities of Hubble. Radio telescopes will give similar precision at longer wavelengths, and a new space telescope (the Next Generation Space Telescope) will be launched and will probe yet deeper into the cosmos, and so back further in time, to when galaxies were first forming. A new generation of terrestrial optical telescopes will be built. The University of California and the California Institute of Technology (Caltech) have teamed up to design and build a telescope with a 30-meter diameter primary mirror, dubbed the California Extremely Large Telescope (CELT). And as though that was not big enough, the European Southern Observatory has plans for a 100-meter telescope: OWL, for OverWhelmingly Large (or perhaps Observatory at World Level). Infrared observations will contribute also to the growing accumulation of data on protogalaxies, so that by 2020 we will have as good an understanding of the formation of galaxies and of their evolution as we do now of the stars within them.

So much for the kind of physics conducted, of necessity, by large international collaborations and planned decades in advance. It is much more difficult to guess what will be discovered in the areas of physics where the scale of phenomena falls somewhere between the macrocosm of astrophysics and cosmology and the microcosm of high-energy particle physics. I expect that the rapid progress in quantum optics and the area of physics centered on the manipulation and control of individual atoms and molecules will continue. The spin-offs from these developments will continue to make their way to the marketplace in fields as diverse as information technology and telecommunications on the one hand and genetic engineering and pharmacology on the other. By 2010 a rudimentary quantum computer will have been developed. Nanometer-scale electronic devices will be used in digital computers, pushing Moore's law to its limits. Microelectrical mechanical systems (MEMS) will be emerging from research laboratories into manufacturing, as the dream of molecular machinery is translated into practical reality.

Much less likely is the discovery of a room-temperature superconductor, but that would have such far-reaching technological consequences that the pressure to find one is very intense. But I would be disappointed if we continue to search much longer with no satisfactory theory of the high-critical-temperature superconductors to guide us.

I guess that controlled thermonuclear power will remain an attractive mirage on the horizon of discovery for at least the first twenty years of the century. But there are many compelling reasons to search for alternative sources of usable energy, not least the pollution and environmental damage our present use of fossil fuel generates. The clamor for cheap energy must be addressed, and even if the major industrial powers can curb their profligate waste, the legitimate needs of the developing world must still be met. The politicians may at last commit themselves to the right ends, but it will be science that provides the means, and our existing science is inadequate. What's more, we have to solve this problem soon: we have more and more people and less and less fuel.

Most of physics research is conducted in fields of no great glamour, where the problems are tough and long-standing, and advances are incremental. Some impression of the range and diversity of this activity is suggested by the areas covered at international conferences: a recent list of more than fifty, held in just one month, included (in no particular order) purification and crystallization of proteins, radioactive materials, uranium fuel, high-magnetic fields, transport in disordered systems, microprobe technology, reactor safety, silicon devices, diamond technology, fusion energy, complex fluids and bio-physics, therapeutic radiology, complexity and fractals, electrical insulation and dielectric phenomena, signal and image processing, as well as meetings on less focused topics. A hundred years ago it was possible for an active research physicist to be informed about all that was going on in physics; obviously, no one physicist could achieve this today. The work reported at such conferences is not unexciting, and much of it is likely to lead to appli-cations of benefit to society. But it is not headline-grabbing material. It will not win Nobel Prizes. Like all physicists today, I am ignorant of much of what my fellow physicists do, and hope that you, and they, will excuse the unevenness of my survey.

I've gone out on a limb in hazarding these guesses, and by the time you read this I may already be proved wrong. The fundamental bases of physics seem to me to be secure, and I do not expect to see them undermined. But I would be very excited if they were! Of one thing I am certain: there *will* be surprises. There always are.

Notes

★ ★ ★

CHAPTER 1

1. He told his son, "Today I have made a discovery which is as important as Newton's discovery."

2. A hundred years ago, low-temperature research was able to approach a few degrees above absolute zero; experiments now explore temperatures a few billionths of a degree above absolute zero.

3. The BCS theory, named for Bardeen, Cooper, and Schrieffer, who were to receive the 1972 Nobel Prize for their work.

4. Nitrogen liquefies at $-196°$ C, equivalent to 77 K, that is to say 77 degrees above absolute zero; helium remains a gas down to 4.2 K.

CHAPTER 2

1. Thomson had at age twenty-eight been appointed as Cavendish Professor at the University of Cambridge, a position previously occupied by Maxwell. He said of his discovery of the electron, "Could anything at first sight seem more impractical than a body which is so small that its mass is an insignificant fraction of the mass of an atom of hydrogen?"

2. The thesis was entitled "On a New Determination of Molecular Dimensions." He had submitted a doctoral thesis on a different topic to the University of Zürich in 1901, but it was not accepted.

3. Einstein received the 1921 Nobel Prize for Physics, but not for his relativity theory. It was for his discovery of the law of the photoelectric effect. First noticed by Heinrich Hertz in the course of his experiments on the production of radio waves, the photoelectric effect is the emission of electrically charged particles from the surface of metals when illuminated by light. In 1900 Philipp Lenard showed that these particles were the same as the electrons so recently discovered by Thomson. Einstein's explanation introduced the idea that light could be regarded as behaving like particles—what we now call *photons*—and this idea is one of the basic ingredients of quantum theory. Friedrich Ostwald nominated Einstein in 1910, but it was more than a decade before he received the prize.

4. The amp is the unit of electric current, the ohm that of electrical resistance, and the volt that of electrical potential.

5. It is not only visible light that is radiated; there is also emission at frequencies below that of red light (infrared) and above that of violet light (ultraviolet). The spectrum extends indefinitely in both directions.

6. Figure 3.6 shows the curve plotting this function, based on experimental data. The shape of the left-hand portion of that curve had already been deduced by Lord Rayleigh (who followed Maxwell as Cavendish Professor at Cambridge and was in turn succeeded by J. J. Thomson) and James Jeans.

7. Rutherford's experiments and theoretical insights played a fundamental role in the development of physics in the twentieth century. He discovered the atomic nucleus, predicted the existence of the neutron, and, as was recorded in his citation for a Nobel Prize in 1908, he made "investigations into the disintegration of the elements, and the chemistry of radioactive substances." But his prize was not for physics; it was for chemistry!

8. Rowland was one of the founders of the American Physical Society in 1899 and its first president. In his address at its first meeting, he asked questions which still today define the scope of fundamental physics: "What is matter; what is gravitation; what is ether and the radiation through it; what is electricity and magnetism; how are these connected, and what is their relation to heat? These are the great problems of the universe." He made important contributions to research in electromagnetism and in thermal physics before his development of the concave spectral gratings (obviating the need for a lens) for which he is best known.

9. Hamilton was a most remarkable man, a friend of the poet Wordsworth and an accomplished linguist (he had mastered Latin, Greek, and Hebrew by the age of five and was later able to speak thirteen "dead" languages). He was appointed Professor of

Astronomy at Trinity College, Dublin, while still an undergraduate and only twenty-one years old and is also celebrated as the inventor of *quaternions* (a generalization of complex numbers with three imaginary units). He was so taken with this discovery that he carved the defining formulas: $i^2 = j^2 = k^2 = i\,j\,k = -1$ in the stone of Brougham Bridge while out walking with his wife. And later wrote, perhaps immodestly, "I still must assert that this discovery appears to me to be as important for the middle of the nineteenth century as the discovery of fluxions [the calculus] was for the close of the seventeenth."

10. A maximum or a minimum is an extremum.

11. This has a close analogue in the definition of a straight line as the shortest line joining two points, only here the "points" are specifications of the initial and final configurations of the system under consideration, and "distance" is defined in an abstract (but precise) way.

CHAPTER 3

1. In a report from the Harvard College Observatory in 1912 appears a "statement regarding the periods of 25 variable stars in the Small Magellanic Cloud [. . .] prepared by Miss [Henrietta Swan] Leavitt." In it she emphasized that "A remarkable relation between the brightness of these variables and the length of their periods will be noticed."

2. The *exact* value is 2.99792458×10^8 meters per second. In 1983 it was decided to fix the standard meter so that this was so.

3. Hubble's original determination was off by a factor of more than two: he did not know that there are two distinct populations of Cepheid variables. This was discovered by Walter Baade, whose observations of stars in the center of the Andromeda Galaxy was facilitated by the blackout in the area around Mount Wilson during World War II. He later used the 200-inch Mount Palomar telescope to confirm the need for adjustment to Hubble's constant. In fact, there was a further source of error in Hubble's original determination: he confused distant star clusters with individual stars. Nevertheless, his observations and the law which bears his name helped lay the foundation for all of modern astrophysics and cosmology.

4. The Hubble expansion rate allows one to estimate the age of the universe. From the rate one can determine a characteristic time—the Hubble time—which on the basis of recent observations is around 15 billion years. The relationship between this time and the age of the universe is not straightforward, however; it depends on factors like the average density of matter in the universe. Lively controversy still

surrounds the details of cosmic origins, but the consensus yields an age for the universe of around 13 billion years—comparable with the age of the oldest stars.

5. I cannot believe that our star, our sun, is so special that it alone has an orbiting planet with conditions comfortable enough to support some form of life. Life on Earth is extraordinarily tenacious and widespread; there are living creatures in the depths of the oceans, deep in the rocks, in the searing temperatures of the fumaroles of volcanoes, in glaciers and ice floes. Why, then, should we assume that life cannot arise in a huge variety of environments? It is more than likely that life forms elsewhere in the universe differ from those on Earth, perhaps more than a zebra from an apple tree, or a bacillus from a baboon. But I am utterly convinced that life exists in untold variations elsewhere in the universe.

6. Other giant telescopes include the twin Keck telescopes on Mauna Kea in Hawaii, each of which has a segmented mirror four times the area of the 200-inch Hale Telescope on Mount Palomar. In Chile, a consortium of eighteen U.K. universities will build VISTA—the Visible and Infrared Survey Telescope for Astronomy—a 4-meter survey telescope (designed to have a wide field of view).

7. We will have more to say about black holes later in this chapter and also in chapter 9.

8. Not entirely in jest, Hewish and Bell labeled the signals LGM, for Little Green Men, since they thought that such regularity might indicate an intelligent source.

9. The different endpoints for stellar evolution can be explained by astrophysics. Some stars simply fade like cooling cinders when they have exhausted all the thermonuclear reactions which made them shine. Other, more massive stars collapse under their own weight when their fuel is exhausted, the radiation which had previously staved off this collapse having faded. The extent to which this collapse continues depends on their mass in a subtle way which involves quantum mechanics. If the star is not much more massive than the sun, it will eventually become a white dwarf, usually having first gone through a stage in which it expanded into a "red giant." The eventual fate of a star with a mass between about 1.5 and 3 times the mass of the sun, is to become a neutron star, perhaps a pulsar. But if, at the end of its life, its mass exceeds three solar masses, nothing can halt its continuing collapse, and it becomes a black hole.

10. This was the brightest supernova visible from Earth since 1054, when the source of the Crab Nebula was observed. There is an excellent account of this supernova in Laurence Marshall, *The Supernova Story* (Princeton: Princeton University Press, 1994).

11. They were told what it was by Robert Dicke, from nearby Princeton University.

12. Physicists use a temperature scale, the Kelvin scale, related to the more famil-

iar Celsius scale, which takes its zero the freezing point of water. But the zero on the Kelvin scale is the absolute zero of thermodynamics, which corresponds to $-273°C$.

13. At high temperatures atoms break up into charged ions and free electrons, forming what is called a plasma. The plasma in the early universe was opaque to light.

14. Called after Chandrasekhar, who was always affectionately known by this abbreviated version of his name.

CHAPTER 4

1. "Given for one instant an intelligence which could comprehend all the forces by which nature is animated and the respective positions of the beings which compose it, if moreover, this intelligence were vast enough to submit these data to analysis, it would embrace in the same formula both the movements of the largest bodies in the universe and those of the lightest atom; to it nothing would be uncertain, and the future as the past would be present to its eyes." *Oeuvres*, vol. 7, *Théorie Analytique de Probabilité* (1812–1820), Introduction.

2. To give a simple example, if the process under discussion is the fall of a stone dropped from rest, the action is obtained by multiplying together the weight of the stone, the height from which it was dropped, and the time it took to reach the ground.

3. So, for example, the continuous flow of time can be represented in an approximate fashion by the count of the discrete, passing seconds on a digital watch.

4. The simple stone which marks Planck's grave in Göttingen bears only his name and the inscription "$h = 6,62 \cdot 10^{-27}$ erg@sec." (1 erg $= 10^{-7}$ joule.) A measurement made in 1998 at the National Institute of Standards and Technology in Gaithersburg, Maryland, found h to be $6.62606891 \times 10^{-34}$ joule second, with an accuracy of 89 parts per billion. (For comparison, the action of a car traveling for a second at 60 mph is about a million joule second.)

5. *The age of extremes: the short twentieth century, 1914–1991.* (London: Michael Joseph, 1994).

6. He was also to become a virulent anti-Semite and a prominent member of the Nazi Party. He was particularly incensed by the acclaim earned by "the Jew Einstein" for the theory of relativity. In his autobiography he wrote that relativity "was a fraud, which one could have suspected from the first with more racial knowledge than was then disseminated, since its originator Einstein was a Jew." (*Erinnerungen*, unpublished typescript (1943), as quoted in Alan D. Beyerohen: *Scientists under Hitler* (New Haven: Yale University Press, 1977)). In 1925 a paper by R. N. Ghosh, an

Indian scientist writing in English, was published in *Zeitschrift für Physik* (the flagship publication of the German Physical Society), without even being translated into German. This so provoked Lenard's xenophobia that he posted a notice on his door: "No entry to Jews and members of the German Physical Society."

7. "In sum, one can say that there is hardly one among the great problems in which modern physics is so rich to which Einstein has not made a remarkable contribution. That he may sometimes have missed the target in his speculations, as, for example, in his hypothesis of light-quanta, cannot really be held much against him, for it is not possible to introduce really new ideas even in the most exact sciences without sometimes taking a risk." Statement supporting the nomination of Einstein for membership in the Prussian Academy of Sciences, June 12, 1913; signed by four distinguished German physicists, including Planck.

8. So, returning to our example of a falling stone, if its mass was 1 kilogram and it fell from a height of 5 meters, the action would be 10^{35} times h!

9. For example, although it is not possible to determine just when any single radioactive atom will decay, it *is* possible to state with a very high degree of accuracy that after twenty-nine years one-half of the very large number of atoms in a macroscopic sample of radioactive strontium will have decayed.

10. Quantum mechanics was devised in 1925 by Heisenberg and Schrödinger, working independently; and surprisingly, their formulations seemed at first to be quite different. But they were quickly shown to be mathematically equivalent, even though they seem to be very unlike one another. Schrödinger's *wave mechanics* and Heisenberg's *matrix mechanics* thus yield the same predictions for all applications; one may choose whichever is the more convenient for the problem at hand.

11. When two waves remain "in step" so that successive crests of each occur always in time with one another, they are said to be coherent. Coherence is a prerequisite for the characteristic wave phenomenon of interference, which occurs when two waves reinforce or cancel out where they are superposed or cross one another. Decoherence occurs when the waves either drift out of step, or are jolted out of step by some outside intervention. It is only when the quantum wave-functions of two states are coherent that it is meaningful to talk of a quantum superposition of those states.

12. Aspect's first experiments on the Bell inequalities were published in 1982. His and other research groups have made further experiments which continue to refine and substantiate the results.

13. If the two values of a bit (0 or 1) in a conventional computer are represented by two distinct states of an atom, a coherent superposition of those two states will represent *both* values of the bit at the same time. This is called a *qubit*. By working

with qubits, a quantum computer will be able to perform many calculations at the same time, allowing an enormous increase in speed. See also chapter 12.

CHAPTER 5

1. Actually there is a more imminent hazard than the earth being flung out of the solar system: the sun itself will eventually swell up to engulf the earth!

2. "Big whirls have little whirls / That feed on their velocity; / And little whirls have lesser whirls, / And so on to viscosity." Lewis Richardson, in his 1920 paper "The Supply of Energy from and to Atmospheric Eddies."

CHAPTER 6

1. An advertisement for the Eurostar train, which travels from London to Paris through the Channel Tunnel, reads, "As if by magic Paris arrived"—so perhaps this relativistic way of looking at things is catching on!

2. For the curious, it's $ds^2 = dx^2 - c^2dt^2$.

3. In Galileo's *Dialogues*, his imaginary character Salviati suggests to Sagredo:

Shut yourself up with some friend in the main cabin below decks on some large ship and have with you there some flies, butterflies, and other small flying animals. Have a large bowl of water with some fish in it; hang up a bottle that empties drop by drop into a wide vessel beneath it. With the ship standing still, observe carefully how the little animals fly with equal speeds to all sides of the cabin. The fish swim indifferently in all directions; the drops fall into the vessel beneath; and, in throwing something to your friend, you need throw it no more strongly in one direction than another, the distances being equal; jumping with your feet together, you pass equal distances in every direction. When you have observed all these things carefully (though there is no doubt that when the ship is standing still everything must happen this way), have the ship proceed with any speed you like, so long as the motion is uniform and not fluctuating this way and that. You will discover not the least change in all the effects named, nor could you tell from any of them whether the ship is moving or standing still.

(Salusbury translation [Chicago: University of Chicago Press, 1953], pp. 199–200)

4. Einstein, however, wrote: "In my own development, Michelson's result has not had a considerable influence. I even do not remember if I knew of it at all when I

wrote my first paper on the subject (1905)." (Letter to F. C. Davenport, February 9, 1954: quoted by A. Pais, in *Subtle is the Lord* (Oxford: Oxford University Press, 1982).

5. Pauli's *exclusion principle* is the key to the quantum-mechanical explanation of atomic structure, and thereby of chemical valency and the Mendeleev periodic table of the elements.

6. The neutron for example has a 50:50 chance of living more than 10.2 minutes before it decays, but many of the other particles live barely long enough to traverse an atomic nucleus, even when traveling at close to the speed of light.

7. Please excuse this one equation: it is the only one (outside of the notes) in this book! Einstein didn't write $E = mc^2$ in 1905, by the way; he used the symbol L for energy. Again in 1912 he wrote $L = mc^2$, but crossed out the L and substituted E.

8. This is rather like the way that a computation of π might yield as successive approximations 3, 3.1, 3.14, 3.142, 3.1416, 3.14159, . . . , each better than the one before. Note that $3.14159 = 3 + 1/10 + 4/100 + 2/1{,}000 - 4/1{,}0000 - 1/100{,}000$, and each term in this expansion is smaller than the preceding one.

9. To continue the analogy with a computation of π. Suppose that the "rules" for the computation gave 3 as the first approximation—which is satisfactory—but that a subsequent calculation, which should have been better, gave $3 + X/10$, with X infinite. This is clearly unacceptable. Suppose, however, that the expression for X is obtained by setting at zero something like p in $1/[p(1 - p)]$, which remains finite so long as p is *not* set at zero. The "subtraction" procedure is then to rewrite $1/[p(1 - p)]$ as $1/p + 1/(1 - p)$, and then to discard the first term before setting p at zero. This procedure would give $X = 1$, which is the "right" answer.

10. The reason we feel the gravitational attraction of the earth—in other words, our weight—but are quite insensible to electrostatic forces is that all the matter in the earth attracts all the matter in our bodies, whereas electric charges attract each other if they are of opposite charge and repel each other if they are of like charge. Both the earth and our bodies are electrically neutral, with positive and negative charges canceling out.

11. The very large number 10^{40} is one which has excited the imagination of some very distinguished scientists, because a similar ratio is found if one divides the age of the universe (for which the best estimates range around 13 billion years) by the time it would take for light to traverse an atomic nucleus. Is this just a coincidence? One comes up with another very large number by estimating the number of protons in the visible universe; this turns out to be close to 10^{80}, the square of 10^{40}. Another coincidence? We do not at present have any theoretic basis for an understanding of these numbers, but it is tempting to suppose that they, and others like them, convey some sort of clue about fundamental physics.

12. Experiments determine a radius for the proton consistent with these ideas: about 10^{-15} meters. For particles like the electron, it is very much smaller, but still not zero.

CHAPTER 7

1. MIT Press, 1965.

2. In "Questions for the Future," his contribution to *The Nature of Matter*, ed. J. H. Mulvey (Oxford University Press, 1981).

3. *Quantum Mechanics*, 3d ed. (Oxford University Press, 1947).

4. Famous also as the founder of modern number theory. His "last theorem," written in the margin of his copy of the *Arithmetica* of Diophantus—"To divide a cube into two other cubes, a fourth power or in general any power whatever into two powers of the same denomination above the second is impossible, and I have assuredly found an admirable proof of this, but the margin is too narrow to contain it"— eluded proof (for Fermat's proof, if he ever really had one, has never been discovered) until 1993, when Andrew Wiles announced that he had a proof. This was at the end of a series of lectures given at the Isaac Newton Institute in Cambridge. But when he came to write up the proof for publication, he discovered a subtle error. It took a further year of intensive effort to fill the loophole, and the resulting tour de force was published in 1995.

5. Its reciprocal—137.0359895—is a number that has fascinated and misled many physicists. At a time when experiments were not yet precise enough to rule it out, Sir Arthur Eddington claimed it was exactly $10^2 + 6^2 + 1$. Other more elaborate formulas have been proposed, none of them convincing, and it remains a challenge for the future to determine its value from a more fundamental theory. Wolfgang Pauli had studied in Sommerfeld's institute in Munich. He regarded the fine structure constant as the link to the "magic-symbolic" world which he had discussed with his friend the psychologist C. G. Jung. Pauli died in the Red Cross Hospital in Zurich; he is said to have been disturbed by the fact that his room number there was 137.

6. A pure imaginary quantity is one which squares to give a *negative* result, unlike a real number the square of which is positive (as W. H. Auden wrote: "Minus times minus is plus / The reason for this we need not discuss"). A complex number is a sum of a real number and an imaginary one. The quantum-mechanical amplitude is in fact complex, and complex numbers play an important role in all the various formulations of quantum mechanics.

7. Minkowski said in 1908: "The views of space and time which I wish to lay

before you have sprung from the soil of experimental physics, and herein lies their strength. They are radical. Henceforth space by itself, and time by itself, are doomed to fade away into mere shadows, and only a kind of union of the two will preserve an independent reality."

8. The formula expressing this link, $S = k \cdot \log W$, is to be found on his tombstone (figure 7.4). In this equation, S is the entropy and W is the probability of a state. The probability is proportional to the number of microstates corresponding to the macrostate in question, for example, the number of different arrangements in space the molecules in a gas may assume. The constant k is called Boltzmann's constant.

9. It was proposed in 1957 by Hugh Everett III in his doctoral dissertation.

CHAPTER 8

1. This is the Pierre Auger Observatory, which consists of two arrays of detectors, one in Utah in the United States and one in Argentina, located so as to allow coverage of the whole sky. Each of these sites will have 1,600 particle detectors deployed over 3,000 square kilometers.

2. The unit is Planck's constant divided by 2π.

3. One such experiment, called K2K, started to collect data in 1999. Neutrinos produced at the synchrotron accelerator operated by the High Energy Accelerator Research Organization (KEK) in the city of Tsukuba are beamed under the "Japanese Alps" to be detected at the Super-Kamiokande detector, which lies 250 kilometers away in the town of Kamioka. The Japanese-Korean-U.S. team running the experiment can then look for any possible effects due to neutrino oscillations by comparing the Super-Kamiokande measurement with the front detector measurement at KEK.

4. A word of caution is appropriate here! Quite apart from possible exotic forms of matter, or even massive neutrinos, compelling evidence from the dynamics of motion both of stars within galaxies and of galaxies themselves within galactic clusters suggests that the mass of visible stars contributes only a small fraction of the total mass of matter in the universe.

5. This from the Greek *leptos*, small, since the electron and the muon have smaller mass than the proton.

6. From the Greek *hadros*, thick, bulky.

7. Murray Gell-Mann took the name *quark* from the verse that opens book 2, chapter 4, of *Finnegans Wake*: "——Three quarks for Muster Mark!" The pronunciation usually rhymes with "park" in Europe, but with "pork" in the United States. In a

private letter to the editor of the *Oxford English Dictionary* (27 June 1978), Gell-Mann wrote: "I employed the sound 'quork' for several weeks in 1963 before noticing 'quark' in 'Finnegans Wake', which I had perused from time to time since it appeared in 1939 . . . I needed an excuse for retaining the pronunciation quork despite the occurrence of 'Mark,' 'bark,' 'mark,' and so forth in Finnegans Wake. I found that excuse by supposing that one ingredient of the line 'Three quarks for Muster Mark' was a cry of 'Three quarts for Mister . . .' heard in H. C. Earwicker's pub."

8. For example, the complex pattern of lines in the spectra of atoms and molecules can be classified according to symmetry principles that reveal details of the underlying physical processes. And the same ideas carry over to the classification and understanding of the elementary particles and their interactions.

9. It is SU(3)⊗SU(2)⊗U(1), where the first factor is associated with the strong interaction and the others with the electroweak. I promise you, this is simpler than it looks!

10. The apparatus they had used was originally designed for a quite different experiment—to measure the decays of a transient state of two pions—and they made brilliant adaptations for their purpose.

CHAPTER 9

1. Physicists use *acceleration* for any change in velocity, either a change in speed or direction.

2. A. Einstein, *The Meaning of Relativity* (Princeton: Princeton University Press, 1950), p. 59.

3. He had also tried to determine the speed of light by signaling with lanterns between neighboring peaks in the Hartz mountains!

4. *Space, Time and Gravitation: An Outline of the General Relativity Theory* (Cambridge: Cambridge University Press, 1920).

CHAPTER 10

1. From the experimental perspective, the quarks and leptons of the standard model have no apparent structure, even at the smallest scales probed. They are pointlike even at the level of 10^{-19} meters.

2. I like this way of characterizing a theory that has very little scope for arbitrary adjustment and I borrow it from Steven Weinberg, with acknowledgment. He uses it

in his excellent book *Dreams of a Final Theory: The Search for the Fundamental Laws of Nature* (Hutchinson Radius, 1993).

3. More precisely, compactification on a circle of radius R is equivalent to compactification on a circle of radius l_P^2/R, where l_P = the Planck length.

4. This is another one of the intriguing "large numbers": $(10^{60})^2 = (10^{40})^3$.

CHAPTER 11

1. More precisely, the quantity I will refer to as the size of the universe is a scale factor, which has meaning even for an open, infinite universe.

2. "Some say the world will end in fire, / Some say in ice." From "Fire and Ice," in *The Poetry of Robert Frost* (New York: Holt, Rinehart and Winston, 1969).

3. What is meant by curvature of the universe? A straight line may be defined as the shortest line between two points, and then a triangle is made up of three straight lines. In Euclidean geometry, the angles of a triangle always add up to 180°, but this need not be true. For example, on the surface of a sphere (with the edges of the triangle all lying in the surface, like great circles on the surface of the earth), the angles add up to more than 180°, while on the surface of a saddle they add up to less than 180°. A "flat" universe is one with Euclidean geometry. In Euclidean geometry, the volume enclosed in spheres of larger and larger radii grows in proportion to the cube of the radius; in an open geometry, the volume enclosed increases faster than this, and in closed geometry, more slowly.

4. Sakharov played a major role in the development of the Soviet Union's H-bomb. He later became the leading figure in the civil rights movement in the USSR, and was awarded the Nobel Peace Prize in 1975. His outspokenness brought him into increasing conflict with the government, and when he opposed the war in Afghanistan, he was banished to Gorky (now returned to its former name of Nizhni Novgorod) and denied contact with other scientists. An international campaign to end his persecution only succeeded after seven years, when President Gorbachev, in a phone call, invited him to return to Moscow and "go back to your patriotic work."

5. This comes from observations at the Kamiokande facility. The experiments are located in the Kamioka Mining and Smelting Company's Mozumi Mine, and Kamiokande was originally designed as a nucleon decay experiment. Another important observation at Kamioka was the detection of neutrinos from the 1987A supernova, heralding a new kind of astronomy. It was the upgraded detector, Super-Kamiokande, which found evidence for the neutrino mass.

CHAPTER 12

1. Chemists might bridle at the unfortunate (and indeed uncharacteristic) immodesty of Paul Dirac, who wrote, à propos of quantum mechanics, that "The underlying physical laws necessary for the mathematical theory of a large part of physics and the whole of chemistry are thus completely known, and the difficulty is only that the exact application of these laws leads to equations much too complicated to be soluble." *Proceedings of the Royal Society of London*, ser. A, (1929): 714. But it is true that no new laws of *physics* are needed to explain all of chemistry now that we have formulated the basic laws of quantum mechanics.

2. At times politicians have found it difficult to separate the categories. "Communism is Soviet power plus the electrification of the whole country." V. I. Lenin, Report to the 8th Congress of the Communist Party (1920).

3. According to R. Stanley Williams of Hewlett Packard Laboratories, "The amount of information that people will be able to carry with them as back-up memory [using nanotechnology] will be equivalent to the contents of every book ever written." *Physics World* December 1999, p. 51.

4. *Laser* is an acronym of "light amplification by the stimulated emission of radiation." A microwave analogue (a maser) had been demonstrated earlier, in 1954.

5. The effect exploited in the ring-laser gyroscope is named after Georges Sagnac who measured interference in a rotating interferometer in 1913, but the idea to use an interferometer to detect the rotation of the earth goes back to 1893, to Oliver Lodge and to Joseph Larmor—whom we met in chapter 2. Pauli said of the Sagnac effect that it is essentially the optical analogue of the Foucault pendulum. And, as with the Foucault pendulum, there is a latitude-dependent factor that results in a similar ring laser in Germany measuring a different frequency.

6. These are based on a phenomenon studied more than a century ago by Pierre Curie. Piezoelectric materials change their shape when subjected to an electric voltage and, conversely, generate an electric voltage when squeezed. Piezoelectricity produces the sparks often used to light gas cookers and is also exploited in the training shoes that generate flashing red lights when you tread heavily enough.

CHAPTER 13

1. In the last months of 2000, experiments at CERN's LEP collider produced some evidence for the Higgs particle, but not sufficiently strong to be conclusive. The CERN management agonized over whether to allow data taking to continue for

another year, but decided against, because it would have led to a postponement of the planned shutdown of LEP to make its tunnel available for the construction of the LHC. They extended the LEP program for one month, but the additional data collected was still not sufficient to overcome the caution of the experimenters. They reluctantly had to see their lead in the race to discover the Higgs particle pass from Europe to the United States, where Fermilab is now the front runner.

2. For example, at CERN and at the RHIC (relativistic heavy-ion collider) at Brookhaven National Laboratory on Long Island.

3. There is a similar long baseline experiment in which a neutrino beam from CERN will be directed under the Alps toward a detector 732 kilometers away, in an underground laboratory beside the Gran Sasso Tunnel on the highway connecting Teramo to Rome. And the K2K experiment mentioned in chapter 8 has already borne fruit: neutrinos produced at the KEK accelerator have been detected at Kamioka. A serious accident on November 12, 2001, destroyed more than half of the photomultiplier tubes in the Super-Kamiokande detector, but they will be replaced to allow the experiment to continue.

4. Aristotle, in his book *On the Heavens*, added a fifth element to the four (earth, air, fire, and water) traditionally used to describe matter in the sphere below that of the moon. This was the incorruptible substance of the celestial realm, which he named *quintessence*.

Glossary

★ ★ ★

Action: A fundamental quantity in mechanics from which can be derived the motion of a dynamical system. It plays a central role in the analytic approach to classical mechanics and is likewise of importance in quantum mechanics. Planck's constant is associated with the quantization of action.

Antiparticle: Relativistic quantum field theory predicts that to every species of particle there corresponds a dual kind of particle having the same mass but opposite electric charge (and also opposite values for other chargelike properties). For example, the electron and the positron are particle and antiparticle to each other. Some kinds of neutral particle are identical to their antiparticles; for example, the photon.

Atom: The smallest unit of a chemical element. An atom has a nucleus composed of protons and neutrons, surrounded by electrons whose arrangement determines the chemical properties of the atom.

Big bang: The cataclysmic event, believed to have occurred some 10^{10} years ago, which gave birth to the universe.

Black hole: A region of space from which general relativity theory demands that nothing can emerge, because of overwhelming gravitational attraction. A possible outcome of stellar evolution for sufficiently massive stars.

Black-body radiation: The electromagnetic radiation emitted by a completely black object; the spectrum of the radiation depends only on the temperature of the object.

Chaos: The area of physics that deals with nonlinear systems exhibiting extreme sensitivity to initial conditions.

Classical physics: Physics as understood before the advent of quantum theory and Einstein's theory of relativity.

Coherence: Property ascribed to waves that are "in step" with one another, so that their troughs and crests coincide in space and time.

Cosmic microwave background radiation: The thermal black-body radiation which perfuses the universe, originating from the big bang and now at a temperature of 2.7 K.

Cosmology: The study of the overall structure of the universe, its origin and evolution.

Cyclotron: A device for generating a beam of elementary particles. They are made to follow a spiral path between the poles of a magnet along which they are accelerated by an alternating high-voltage electric field.

Electrodynamics: The interacting behavior of electric charges and the electromagnetic field.

Electromagnetism: The combination of electrical and magnetic phenomena, first brought together by Maxwell in his electromagnetic theory.

Electron: One of the fundamental particles and the first to be discovered; a constituent of all atoms. The number of electrons, which are negatively charged, needed to balance the postive charge of the nucleus of an atom determines the chemical properties of the atom.

Energy: The capacity to do work.

Entropy: A thermodynamic quantity related to the degree of disorder in a macroscopic state.

Field: A physical property extending through space and time.

Galaxy: A congregation of stars, typically around 10^{11} in number.

Gamma rays: Electromagnetic radiation of extremely high frequency, beyond that of x rays. Gamma rays are typically associated with quantum transitions between different energy states of an atomic nucleus but can also be generated by other high-energy processes.

Hydrogen: The element with the simplest atomic structure: a single electron bound to a nucleus consisting of a single proton.

Gravity: A force of attraction experienced universally between all massive entities. The most familiar of the fundamental forces, the gravitational force attracting us toward the earth is what we experience as our weight.

Interference: The phenomenon whereby waves passing through one another either augment one another (constructive interference, where troughs coincide with troughs and crests with crests) or weaken each other (destructive interference, where troughs coincide with crests).

Interferometer: A device which exploits interference phenomena to measure small displacements. It is also used to make precise measurements of wavelength.

Laser: A device that produces light through the action of many atoms acting coherently rather than independently. Light from a laser is typically monochromatic (i.e., it has a sharply defined wavelength) and may be very intense.

Matrix mechanics: Heisenberg's formulation of quantum mechanics, in which the mathematics is related to that of matrices.

Molecule: A chemically bound group of atoms.

Neutron: One of the particles that constitute atomic nuclei. It is electrically neutral.

Neutron star: A collapsed star, largely composed of neutrons. It has a size similar to that of Earth but a mass like that of the sun, so its density is enormous, comparable to that of an atomic nucleus.

Phase space: An abstract space, points in which are identified with the state of a dynamical system.

Photon: The particlelike quantum of electromagnetic radiation.

Planck's constant: One of the fundamental constants, characteristic of quantum mechanics, introduced by Max Planck in his explanation of the spectrum of black-body radiation. It determines the scale of quantum phenomena.

Positron: The antiparticle of the electron.

Proton: One of the particles that constitute atomic nuclei. It carries a single unit of positive charge. The nucleus of a hydrogen atom is a proton.

Pulsar: A star observed as a regularly pulsating object. Pulsars are believed to be rotating neutron stars.

Quantum: A discrete quantity of energy—an elementary excitation of a field.

Quantum mechanics: The general laws underlying the behavior of physical systems. The laws of quantum mechanics deviate significantly from those of classical mechanics in explaining phenomena at very small scales, such as those of the atom.

Quasar: Acronym for quasi-stellar radio source: brilliant astronomical sources of radio waves which are so compact as to appear pointlike, as do stars. They are believed to be intensely active distant galaxies.

Radioactivity: The process whereby an atomic nucleus spontaneously emits an energetic particle and so becomes the nucleus of a chemically different kind of atom.

Redshift: Displacement of characteristic lines in the spectrum toward longer wavelengths. It is associated with recession of the source relative to the detector.

Reductionism: The scientific method that seeks to explain complex phenomena in terms of simpler, more general underlying laws.

Relativity: The theory that describes how different observers represent the same events with respect to their different frames of reference. Galilean relativity refers to the pre-Einsteinian theory for observers moving at constant speed in a straight line relative to one another. Einstein's special theory modifies that to make it consistent with the universal constancy of the speed of light that is required by Maxwell's theory of electromagnetism. His general theory extends this to observers in relative accelerated motion and also incorporates a theory of gravitation.

Spacetime: The union of space with time in Einstein's relativity theory.

Spectrum: Range of frequencies of radiation observed in a physical process. The spectrum of electromagnetic radiation extends in both directions beyond the visible frequencies of the colors of the rainbow: beneath the low frequency of red light and above the high frequency of violet. Below the red comes infrared and then microwaves and the radiation used for television and radio transmission. At the other extreme, with frequencies higher than those of visible light, come ultraviolet, x rays, and gamma rays. Because quantum mechanics relates frequency to energy, one may also refer to an energy spectrum.

String theory: A theory in which the elementary constituents of matter are regarded as extended objects, like strings, rather than pointlike.

Superconductivity: The phenomenon whereby at low temperature certain materials completely lose their resistance to the passage of electric current.

Thermodynamics: Theory of the transfer of heat. The theory deals largely with the collision and interaction of particles and radiation in large near-equilibrium systems.

Thermonuclear fusion: The joining together of atomic nuclei (usually hydrogen or other light nuclei) when they collide at temperatures such as occur in the center of stars. Mass lost in the process is converted into energy.

Transistor: A device exploiting the quantum-mechanical properties of certain materials to control electronic processes.

Trajectory: A path through space followed by a projectile. More generally, the path of the point representing the configuration of a dynamical system as it evolves through phase space.

Ultrasound: Waves like sound waves, but with a much higher frequency putting them outside the range of human audio perception.

Wave mechanics: Schrödinger's formulation of quantum mechanics, in which the mathematics is related to that of waves.

Wavelength: The distance between successive crests in a train of waves of well-defined frequency.

Waves: The term used in physics to describe something that repeats regularly in space and time. Wave phenomena are to be found in almost every branch of physics.

White dwarf: A collapsed stellar remnant—"white" because it is still glowing white-hot; "dwarf" because, though still as massive as a star that has not yet exhausted its thermonuclear fuel, it is many orders of magnitude smaller.

X rays: Electromagnetic radiation with a wavelength much shorter than that of ultraviolet light.

Suggestions for Further Reading

★ ★ ★

Abbott, Edwin A. *Flatland: A Romance of Many Dimensions*. Mineola, N.Y.: Dover Books, 1992.

Allday, J. *Quarks, Leptons and the Big Bang*. Bristol, U.K.: Institute of Physics Publishing, 1997.

Ball, Philip. *Made to Measure: New Materials for the 21st Century*. Princeton, N.J.: Princeton University Press, 1999.

Blair, David, and Geoff McNamara. *Ripples on a Cosmic Sea: The Search for Gravitational Waves*. Reading, Mass.: Perseus Books, 1999.

Close, Frank. *Lucifer's Legacy: The Meaning of Asymmetry*. Oxford: Oxford University Press, 2000.

Cohen, Jack, and Ian Stewart. *The Collapse of Chaos: Discovering Simplicity in a Complex World*. New York: Penguin Books, 1995.

Ferris, Timothy. *The Whole Shebang: A State-of-the-Universe(s) Report*. New York: Touchstone Books, 1998.

Feynman, Richard P. *QED: The Strange Theory of Light and Matter*. Princeton, N.J.: Princeton University Press, 1988.

Fowler, T. Kenneth. *The Fusion Quest*. Baltimore, Md.: Johns Hopkins University Press, 1997.

Fraser, Gordon. *Antimatter—the Ultimate Mirror*. Cambridge: Cambridge University Press, 2000.

Fritzsch, Harald. *An Equation That Changed the World: Newton, Einstein and the Theory of Relativity*. Chicago: University of Chicago Press, 1997.

Gamow, George. *Mr. Tompkins in Paperback*. Cambridge: Cambridge University Press, 1993.

Greene, Brian. *The Elegant Universe: Superstrings, Hidden Dimensions, and the Quest for the Ultimate Theory*. New York: W. W. Norton and Company, 1999.

Gribbin, John R. *Case of the Missing Neutrinos: And Other Curious Phenomena of the Universe*. New York: Fromm International, 1998.

Gribbin, John R., et al. *Q Is for Quantum: An Encyclopedia of Particle Physics*. New York: Free Press, 1999.

Gribbin, John R., and Simon Goodwin. *Origins: Our Place in Hubble's Universe*. Woodstock, N.Y.: Overlook Press, 1998.

Gribbin, John, and Mark Chimsky, eds. *Schrödinger's Kittens and the Search for Reality*. New York: Little, Brown and Company, 1996.

Gross, Michael. *Travels to the Nanoworld: Miniature Machinery in Nature and Technology*. New York: Plenum Press, 1999.

Guth, Alan H., and Alan H. Lightman. *The Inflationary Universe: The Quest for a New Theory of Cosmic Origins*. Reading, Mass.: Perseus Books, 1998.

Hall, Nina, ed. *Exploring Chaos: A Guide to the New Science of Disorder*. New York: W. W. Norton and Company, 1993.

Hawking, Stephen. *A Brief History of Time: The Updated and Expanded Tenth Anniversary Edition*. New York: Bantam Books, 1998.

Hoskin, Michael, ed. *The Cambridge Illustrated History of Astronomy*. Cambridge: Cambridge University Press, 1997.

Kaku, Michio. *Hyperspace: A Scientific Odyssey through Parallel Universes, Time Warps and the Tenth Dimension*. New York: Anchor Books, 1995.

———. *Visions: How Science Will Revolutionize the 21st Century*. New York: Bantam Books, 1998.

Kane, Gordon, and Heather Mimnaugh. *The Particle Garden: Our Universe as Understood by Particle Physicists*. Reading, Mass.: Perseus Books, 1996.

Lindley, David. *Where Does the Weirdness Go? Why Quantum Mechanics Is Strange, but Not as Strange as You Think*. New York: HarperCollins, 1997.

Morrison, Philip, Phyllis Morrison, Charles Eames and Ray Eames. *Powers of Ten*. New York: W. H. Freeman and Company, 1995.

Rees, Martin. *Before the Beginning: Our Universe and Others*. Reading, Mass.: Perseus Books, 1998.

Ruelle, David. *Chance and Chaos*. Princeton, N.J.: Princeton University Press, 1993.

Smolin, Lee. *The Life of the Cosmos*. New York: Oxford University Press, 1999.

Smoot, George, and Keay Davidson. *Wrinkles in Time*. New York: Avon Books, 1994.

Treiman, Sam. *The Odd Quantum*. Princeton, N.J.: Princeton University Press, 1999.

Weinberg, Steven. *Dreams of a Final Theory*. New York: Vintage Books, 1994.

———. *The First Three Minutes: A Modern View of the Origin of the Universe*. New York: Basic Books, 1993.

Index of Names

★ ★ ★

General Index

★ ★ ★

LIGO (Laser Interferometer Gravitational Wave Observatory), 132–33, 190

LISA (Laser Interferometer Space Antenna), 133, 178, 190

MACHOs (Massive Compact Halo Objects), 161
magnetic field, 176, 186, 193
matrix mechanics, 86–87, 200
Maxwell's equations, 16, 44, 81
MEMS (MicroElectroMechanical Systems), 182–83, 192
microscopes: atomic force, 184; electron, 183–84; scanning probe, 184
Milky Way, 3, 24, 27–28, 34, 135
missing mass, 160–61, 149, 189, 191
M-theory, 94, 145, 149, 159, 172–73, 189

nanotechnology, 183
neutrinos, 85, 105–8, 118, 156, 161, 189–90; oscillations of, 108, 189
neutron stars, 35–37, 129–30, 134, 189, 198n.9

phase space, 68–69, 71–74, 76, 87
photoelectric effect, 49, 196
photons, 48, 52, 55–62, 83–84, 90–92, 104–5, 110, 112, 118, 124, 145, 156, 167–68, 175–76, 179
plasma, 155–57, 182, 186, 189
Poincaré section, 69–70
polarization, 56–57, 104
positrons, 83, 90–91, 168
pulsars, 36–37, 129; binary pulsar, 128–31

quanta, 47–48, 55, 83–85, 90, 104, 113, 137–38, 142, 200n.7
quantum amplitude, 53–55, 88–91, 94, 96–97, 203n.6
quantum chromodynamics (QCD), 112, 189
quantum computer, 62, 180–81, 192, 200–201n.13

quantum electrodynamics (QED), 8–10, 55, 83–84, 90, 92–93, 105, 110, 112, 173, 182
quantum field theory, 6, 55–56, 82–83, 87, 90, 93, 112, 114, 137, 140, 142–45, 170
quantum jump, 48, 54, 56–57
quantum mechanics, 4–6, 8, 12, 23, 43, 49, 51–64, 82, 85–90, 97, 99, 104, 136–38, 169, 172, 179–81, 198n.9, 200n.10, 207n.1 (ch. 12)
quantum physics, 47, 49, 174
quantum state, 60, 98–99, 182
quarks, 85, 108–13, 118, 138, 152, 155, 167, 189, 204–5n.7
quasars, 35, 127, 147
qubit, 200–201
quintessence, 191

redshift: Doppler, 29, 34, 38, 41, 128–30, 160, 175–76; gravitational, 124–25, 130
reductionism, 9–11, 43, 45
relativity, 5, 12, 17, 77, 80–82, 90, 199n.6; general theory of, 6, 37, 39, 60, 81–82, 94, 96, 119–35, 137–39, 141–42, 146–47, 149–50, 153, 157–58, 161, 190; special theory of, 5, 62, 78–82, 85, 94, 106, 121–22, 138; tests of, 124–31
renormalization, 84, 93, 97, 113, 140, 142

Schrödinger's cat, 58–60, 62, 97–99, 180
Schrödinger's equation, 54, 58, 89, 97–98
solitons, 93, 146, 149
spacetime, 5, 80, 82, 121–22, 124, 127, 137, 139–40, 142, 153, 169
spectra, 20–22, 28–29, 32–34, 41, 46, 125, 128–29, 152, 175–76. *See also* black-body radiation, spectrum of
standard model, 6, 85, 100, 103–4, 112, 114, 116–18, 136–38, 142–43, 155, 161, 172, 189

statistical mechanics, 17, 45–47, 94–97, 103–4, 146, 155

string theory, 5–6, 11, 56, 136–46. *See also* superstrings

strong interaction, 108, 111–12, 117, 138, 167, 189, 205n.9

supercluster, 31–32, 41, 166

superconductors, 7–8, 173–74, 182, 192

superfluid, 182

supergravity, 140

Super-Kamiokande, 106–7, 156, 206n.5

supernovae, 25, 37–39, 152, 164–66, 189

superstrings 117, 139–46, 150, 169–70, 173. *See also* M-theory

supersymmetry, 117, 139–41, 189

symmetry, 5, 84, 94–95, 109–11, 113–15, 136, 140–41, 143–44, 146, 167–68, 189; charge-conjugation (C), 114, 168; parity (P), 113–14; CP, 114, 168, 189, 191; CPT, 114; spontaneous breaking of, 110–11, 114–15, 155, 167–68. *See also* gauge symmetry, supersymmetry

tachyons, 139, 140

thermodynamics, 18, 45–46, 94–96, 147, 155, 187, 198–99n.12

thermonuclear reactions, 36–37, 152, 160–61, 186–87, 192, 198n.9

traps, 62, 86, 175–78, 182

turbulence, 64–66, 68, 75, 96

uncertainty principle, 51–53, 84, 100

unification, 16, 82, 110, 117, 139, 143, 155, 167–68

wave equation, 145. *See also* Schrödinger's equation

wave-function, 54, 57–59, 97, 182, 200n.11

wave mechanics, 54, 86, 200n.10

weak interaction, 105, 110–11, 114–15, 155

white dwarf stars, 36–38, 129, 134, 152

WIMPs (weakly interacting massive particles), 40, 161, 189